On the Integration of Nature

On the Integration of Nature

Post-9/11 Biopolitical Notes

Richard Grossinger

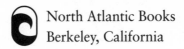

North Atlantic Books
Berkeley, California

Published by
North Atlantic Books
P.O. Box 12327
Berkeley, California 94712

Cover photos: Common Loon of Mount Desert Island by Steve Perrin, World Trade Center behind St. Paul's Chapel by Paula Morrison

Cover photo collage and book design by Paula Morrison
Printed in the United States of America
Distributed to the book trade by Publishers Group West

Some pieces in this book have appeared previously in *House Organ* (Lakewood, Ohio) and *Elixir* (New Lebanon, New York).

On the Integration of Nature: Post-9/11 Biopolitical Notes is sponsored by the Society for the Study of Native Arts and Sciences, a non-profit educational corporation whose goals are to develop an educational and crosscultural perspective linking various scientific, social, and artistic fields; to nurture a holistic view of arts, sciences, humanities, and healing; and to publish and distribute literature on the relationship of mind, body, and nature.

Library of Congress Cataloging-in-Publication Data
Grossinger, Richard, 1944–
    On the integration of nature : post 9-11 biopolitical notes / by Richard Grossinger.
        p. cm.
    Summary: "A series of short essays, stories, and fragments concerning the nature of matter, life, consciousness, and identity in the context of a critique of science and American government policies."—Provided by the publisher.
    ISBN 1-55643-603-3 (pbk.)
    1. Philosophy of nature. 2. Nature—Meditations. I. Title.
BD581.G77 2005
113—dc22

                                                                    2005019681

1   2   3   4   5   6   7   8   9   UNITED   10   09   08   07   06   05

For Pir Zia Inayat Khan, William Kotzwinkle, David Ulansey, Kathy Glass, Charles Stein, Paul Weiss, Joe Richardson, and others unknown to me, who have read my words with intelligence and compassion— thanks.

Table of Contents

Life

Astronomy

Vision

Responses to Dzogchen Buddhist Commentary

In Memory of Friends

Listings in the Table of Contents are more index headings than formal titles (except where duplicated in the text). Like titles, however, they are identified by the first page number of their piece. Some of the shorter items in the book do not appear in the Contents, while a few pieces that fit different topics are "indexed" more than once.

Tomorrow, and tomorrow, and tomorrow,
Creeps in this petty pace from day to day,
To the last syllable of recorded time;
And all our yesterdays have lighted fools
The way to dusty death. Out, out brief candle!
Life's but a walking shadow, a poor player
That struts and frets his hour upon the stage,
And then is heard no more. It is a tale
Told by an Idiot, full of sound and fury,
Signifying nothing.
    —William Shakespeare, *The Tragedy of Macbeth,*
      Act V, Scene 5

On the
Integration
of Nature

At a future time after fossil-fuel depletion—predicating enough energy left then to run an industrial civilization—we might still travel in electric cars, but don't expect to be jetting off to Vegas weekends. There will be no petrol-less planes.

If we were to assemble a network of humongous ships, tens of thousands of them forged in orbit from lunar rock and welded in space by a giant solar mirror; if they were launched (like aircraft out of LAX) in a continually spiraling belt toward Jupiter, adjusting for the shifting positions of the two planets, and then braked, one by one, into Jovian orbit; if a technology were developed to tap the vast methane cloudtops with dangling pipette hoses siphuncling their haul into bulbous tanks; and if the ships, by an unceasing and recalibrating maintenance of their belt, were to deliver the fruits of their culling efficiently back to Earth orbit; if we were to design an orbital pipeline for funnelling Jupiter's stolen gases downward (or were able to offload in space and tugboat the convoy's cargo expediently along a laddered conduit into vast caverns)—our energy supply would be, for all intent and purposes, limitless. The expanding global economy would have all the energy it needed. Everyone on the planet could be a consumer.

It is hard to imagine how to build so many gigantic freighters, keep them on track past Mars through the asteroid belt to Jupiter's domain, how to suck raw gases out of superstorms (an all-but-impossible feat), compress them into metallic vats, and transport them cost-effectively across millions of miles, then transfer them to new vessels or elevators and deliver them from Earth orbit into maintainable subterranean storage prior to refining—to say nothing of the difficulty of lunar mining, the perils of interplanetary flight and close Jovian encounter, the risks of terrorism, *et al.*—but were the entire feat to be executed successfully (and repeatedly), all we would succeed in doing is cooking the Earth and turning it into a sun.

How is that different from what the industrialization of China and India presently portend?

Under a veneer of substance voltaic moons dart in bottomless spools. Science scrapes away curtains to reveal electric-like charges, fields approaching vacuums, beacons at the brim of existence. In order to distinguish photons from light, modern man has blinded himself to the single reality that shapes his destiny, and has placed himself in darkness. Albert Einstein warned, "Once you can accept the universe—as matter expanding into nothing that is something—wearing stripes with plaid comes easy."

At the basis of everything lie apparently indivisible sub-integers, sizeless quantum dots of no internal structure, super-strings of pure energy hundreds of billions of times smaller than the nucleus of an atom.

Matter is naught but penchants given dispositions, fore-runners and fragments of number systems. The materiality of the world is a step above that of a dream—dross of stuff penetrating enchantment of mindedness. Energy and substance are the same thing, substance being simply a mode of concentrating energy that produces illusory palpability of objects. But things are not solidities; they are denser energy bundles, snarls of radiation.

Life is pure space habituated to design, currents kineticized into shapefields. Patterns of energy, porous enough for other patterns to flow through them, rabbits are just thick enough to hold light and color and sensation and the thump of embodiment.

Nature is all keyholes, no keys. Creation does indeed fall off the back of a turtle into a void unimaginable to those who piloted mere threadbare vessels of wood and sails across the Chelonian girth, fearful of a legendary but less dire abyss.

A place is an illusion of objects in a continuity of space/time— a road running from conception to erasure. But—*and this is the key to everything*—the illusion is not itself illusory. While conditional, it is not expendable, cannot be cancelled, because it is driven by the destiny of mindedness and has vested its cocoon in exquisite spreading molecularity. The world's stuff is not hallucination, but a different sort of thing: a karma-driven projection.

Atoms are artifacts and attributes of the same luminosity that, by default, gives us minds. They project secondary reality seamlessly and ingenuously, which provides "being" a domain for its unconscious flow, an outlet for its intrinsic radiance. Otherwise, there would not be a pebble.

It may all be a dream but, once you start to believe, once you join a group dreaming, you are investing in it, building interest every moment. You can't then summarily declare,

"Enough, I don't want to do it anymore. I've read my Buddhist texts. Make my mind stop creating phenomena. Have reality less bright and hard. Show me the bullets are not real. Get me outta here! Wake me up!"

There is no actual dream, so there can be no awakening. Stated differently: once you commit to a landscape, once karma begins to ripen, you can no longer switch from one dream to another. As you experience the scenery as scenery, it becomes scenery. Viewing reality as "unreal" does not cause it to bend a nanometer. You can't push away this sort of specter unless you have something to insert immediately—*immediately*—in its place, an equally viable phenomenology to keep rocks and shadows from sliding right back to where they were.

Why do men and women repair to monasteries as monks and nuns and spend their lives celibate, praying and meditating? Idle question asked by suburbanites, rock stars, bankers, runway models, *et al.*

To try to wake up from the dream. (It is a full-time job.)

To help others awake.

Buddhist meditation is a method of breathing a different scenery, moment by moment, from the mind's innate characteristics, to infiltrate the monopoly of group dreaming. A tiny investment at first, by comparison with the bull-rush of reality, it spreads infinitesimally: another place, another time, another self, a longing for something that has known about us forever, that cannot be fathomed, fleeting before its landscape can be figured or grasped.

Spider, tuning his web, lies at its node, appreciates the silent music of insects drawn by the porch light. Caught by tensility of his lyre, they struggle only to get more entangled and, by then, the arachnid is upon them, its intrinsic absolute mind wrapping them in pellets, extinguishing the billows of their energy fields. Threads are the creature's expression of self, abdominal not cerebral, a motif comprising a history of proto-spiders, archaeo-spiders, and spiders on this planet.

The spider mind is not a separate mind, not a mind separate from its body, though it expresses only one spider's will.

What does not know it exists is (by definition) beyond time and size, invents landscapes of such vastness we could not imagine them. Radiating outward from its unseen center toward its inextricable nature, the spider spells out destiny.

Plants and animals are original entities of electrically active plasm, cohesive in all planes and dimensions, renderable and un. Each is a shell around an eddy, a neuralized configured storm driven by a pump, assembled atop a spout more indelible than thousand-kilometer-deep Jovian cyclones. The character of life is as ineradicable as it is ineffable.

A cell is an intelligent system, as is the transcellular medium it encompasses; so is a frog's kidney; so is Seal Cove Pond, churning up bottom-layer nutrients as autumn chills toward winter, layer by descending layer, currents incrementally sorting and distributing molecules, into October's frost, keeping itself vital, the industrious entities in it alive.

Intelligence does not require ganglia or neural nets, does not even require bodily location. Intelligence is a meta-system that operates as an ocean, a sun, or intergalactically.

A duck floats on Somes Pond, a feathered brown mass.

It is dead. Head and most of a neck hang, skeleton gnawed to a bony string.

When we look to a Martian landscape for life—or even a body of water—we see the dormant turgor of rock, afloat only in our imagination.

Every dog, from the Aztec chihuahua and Euro-dachshund to the eclectic collie, poodle, Great Dane, husky, terrier, spaniel to rottweiler, carries near-identical DNA. Any mutt can smell this from across the street.

How, without a mallet and anvil, have subvisible beads been set along divergent tracks, have breeders altered the topos of mineral itself?

Competition and commensalism are stories we fabricate for a drama that, under microscopy, reveals itself as submolecular motion conducted into cells. The truer principle is not perceptible insofar as our stories are dense enough to blot it out even as they provide its foolproof, irrefutable excuse.

Existence is bare and sheer and doesn't need a story.

While everyone stares into Darwin's populations and their carrying capacities, what they don't see is nature *(gnasciri)*— open-ended motifs of matter under sourceless edict.

The feeling is not reducible to the chemistry of jasmine or the molecules of a breeze.

What began as a trance will end in a trance someday.

Rock is a species of being. The mica-feldspar boulders that rest on the glacial apron of Mount Desert Island, tens if not hundreds of thousands of years, are creatures of another order, cogitating at an unwonted frequency. Their patience makes their motives and movements impalpable to us. Sacred geographer Chris Kaiser named the most inscrutable of these zoids "Wolfie's Rock" in honor of his son.

WR is a complex being, a petrified dinosaur tortoise of so many dimensions that its armatures mesh into separate conclusive (and inconclusive) polygons: trapezoidal segments and warped dodecahedronal spheres.

It is difficult for humans to address WR. His face and eyes are everywhere—and nowhere. Fissures in him house roots of spruces that rustle in breezes. Black moss oozes over one of his foreheads. Lichens crusting his skin feed from his micaceous base.

Where he has perched for epochs, a dank cave gestates under one of his haunches, a crypt dripping rootlets and mud.

This creature is neither smiling nor frowning; it is healing, which includes both frowning and smiling. It is medicinal, but only in the absolute sense.

WR gives up heart-shaped fragments and other arrowhead-like chips not to would-be Mohawk or Passamaquoddy shamans, not to hippie vision-questers who lay chi-sensitive hands on him and petition axial flow, but to chain-smoking, alcoholic penitents whose minds are unclouded by spiritual ambition. Sectors of hide that seem solidly riveted to the core come off like flint in their hands.

Grasped in fingertips, these rock daughters feel vibratory, as they jell faster than the speed of light. While maintaining the hardened shape of their granite stratum, they keep no actual shape at all, simultaneously stone and feather, their qualis shifting as if one were holding very wet clay or tachyonized water.

Like the noncombustion motor of a UFO, Wolfie's Rock bears a slow, weak, but infinitely profound electrical charge.

Too protracted for humans to hear, using neither Indo-European nor Penobscot, the stone speaks its name: Bblllbbb-dddggg. I add our vowels for intonation: Blibdog.

For "world" status, atmosphere is necessary, a cobalt heavens twinkling. Airless rocks like Iapetus and Sedna are not worlds. Mercury, Pluto, even fair Luna are large asteroids. Io is an orbiting volcano.

Callisto and Europa are worlds that harbor oceans under glaciers, habitats for which ice is sky. Titan is a dark town: smoggy, stillborn, tenebrous, and wintry. Neptune, Uranus, Saturn—and other Jovian bodies throughout the universe—

are little more than nitrogen-methane holding tanks.

Climatologists deduce that Venus was formerly temperate; though closer to the Sun, a wet planet. Now a taiga of acidic metal rivers, it draws gases a hundred times more thickly at its surface than Earth—so every breeze is a hurricane. In 900-degree-Fahrenheit nights, rocks glow—solar energy burnishes them by day.

Why something rather than nothing? That haunting seventeenth-century conundrum, even less answerable now, introduces modern philosophy, yet shipwrecks on the reefs of twentieth-century astronomy.

Venus is something—yes—but it has no trees to fall where no one hears them, nor even the bleat of one mouse, a riddle that postpones Leibniz's only by posing another.

Without worlds, stellar wind would roar through a voidless void, shapeless and silent, forever.

Science fails to distinguish between mindedness and a stream of molecules synapsing into gray matter that gives consciousness its proxy and installs consensus reality. The mammalian brain has antecedents in nerve nets of flatworms, ganglia of octopi; these are the streets and boulevards of a future city.

The complexity of wiring in the eye—optic synapse, chiasma, dura, ocular cortex—speaks to the visionary act, a microarchitecture of how the universe torques through substance, deriving form from feedback. As networks emerge out of layered stacks, nodes fabricate functions.

But the eye arises likewise from phenomenology itself. Either way, a brain is flooded with worlds.

Beyond sight is an unexplored thicket, suffused with blobs, sparks, depression waves, spirals, shafts, flashing asterisms, and fading vestiges of after-images, plus the raw material of neural tissue—an onrushing inner sky of immeasurable depth and density.

The eye is joined to the optic cortex as a bulb cultured on it, smack against and *in* it. The cerebral field dances and changes but does not move as the eye moves.

For a moment it falls away entirely. There is a statue of a wooden dog. It is descending toward you, the background unpeeling. The dog comes alive and acts. It changes size and shape; it is a denizen of the brainstem.

Body is not separate from soul, spirit, or psyche. Each is a subtle vibration. Body, while *less* subtle, consolidates psyche and spirit in its timbre. Soul is too subtle for scientists to detect. They do not discern even that it exists, nor do they espy spirit, notwithstanding its footprints all over matter.

Modernism uses the algorithm as its single denominator, the algorithm which is merely the current face of the Oracle that sits at Delphi.

Disincarnate intelligence is the glue of matter, cogent enough not only to conduct molecules but to do so in a way that makes it seem that gravity, heat, and random events exclusively are governing them. Without spirit, gravity would not stick, heat not rouse; the whole cosmodynamic enterprise would fail. Mind is the furnace of matter.

To most indigenous peoples, reality is sustained by ancestors and totems. Objects exist because ghosts impel them.

Subatomic crystalloids mingle into crystals. Clouds, trees, sun, soil are elaborate rubies cultivated on gravity's watch. Rock tempered under his breath, a pagan mage pumps his bellows, exhorting quantum states of hydrogen until carbon and oxygen bubble forth and animalcules slither onto worlds, calculate necessity, spin and soar.

As protein sheets swathe and twist, great rodents and bears shudder to life, adhesion by adhesion, knotting tendrils back and forth around one another and through gaps in buckling layers, placing the without within and out again. In fractal corridors, shear forces squeeze out gazelles, wrens, and other raku that pop free of their matrices. Animals run with a pulsing continuity we recognize as life. Their metabolic engines are nonmechanical. Even in a dream we can tell a bird from a wind-up toy or the shadow of a tumbling leaf.

Alphabets of ginkgo and horseshoe crab pour from boundaries of water and wind; feathered designs become stacks of onyxes and emeralds cascading in Boolean sets. Huddled in tribes, they make networks by which to manage crystalline technologies. They mine oils and ions.

Culture is a creation, flint by flint, bead by bead, pot by pot, growl by vowel.

Life is not an operating system. Life is dimensionless strings of incipient designs realized as progenitors of other ceaselessly differentiating, equally elegant strings—nonlinear algebra of an inherent domain. Where information is lost by chance, enough remains to complete and even enlarge the plan, or another sufficiently beastlike to keep the enterprise going, the wart-hog wheezing.

Our cells are singularly obedient to self-replicating studs that hold meaning together only because there is nothing strong enough yet and persistent enough to prevent it from happening or, once annealed, tear it promptly apart (Picasso: "It's a wonder we don't all melt in our bath"). Vortices maintain lizards and moose, lassoing fey molecules and assembling them in hierarchies, keeping lugubrious stacks from toppling downhill and dispersing in the next breeze. Compact chipmunks squeal and chase each other in and out of the base of a cottage.

Corpses are interrupted states of prior systems. Without a vortex, the cadaver of the woodchuck immediately begins devolving under assault from tiny operating flies and bacteria that steal its molecules and convert them into their own.

A match is struck between oblivion and eternity. That is the only explanation, your one task: follow its light.

Mars' natural geography is interrupted at spots by strutted glassy tubes, translucent tunnels winding and interlocking just beneath the sand, plunging and swerving to avoid each other, casting shadows that follow their shapes, clustering maplike and departmentally around canyons, their riblike bands collapsed here and there catastrophically by massive boulders, their "artificial" surfaces polished enough to beam crisp specular reflections of the sun onto *Mars Global Surveyor*.

Forget Percival Lowell's *canali* mirages bearing water from polar caps to inhabitants of parched cities on a dying world; these are the true Martian pipelines—underpasses, monorails, waterways, septic systems, either a geophysical mirage or a figment of an antediluvian civilization.

There is no gap kinetically between moonlit froth in atmospheres, cream in coffee, and the opacity of immense transplanetary nebulae.

Come morning, mercurial liquors (of a proximal star) drench clouds; a world drinks into grasses, eels, bumper-to-bumper traffic. Orbital velocities meet at the arcs of eggs, uncoiling as chytrid-like, myxozoa-like things.

A disembodied squid, more nuclear than atoms or their nuclei, greater above galaxies than galaxies are above planets, is the single mind that both flatworms and meanings crawl along.

What cannot be probed are the "w's," "q's," and "x's" of this text.

Life has been constructed by signals, gathered as scuzz in pools, then (over a billion or so years of hydraulic stasis) federated, synopsized, and transduced into nonlinear networks. The cell is a machine assembled through hundreds of millions of years of system refinement whereby everything in its entropic mechanism has been mitigated, streamlined to less than zero in order to elude nature's ubiquitous drag. It is such a perfect engine, so much deeper than an algorithm, that not even the most advanced cybernetic technology could have concocted it in less time. Every cumbersome cable and gear has been trimmed and interpolated so tightly that nothing—nothing linear, that is—is left.

That is why entities develop so rapidly, confidently, and seamlessly despite their disruptions of gravity and mass. All the real switches, circuits, and controls have been buried far beneath genetic and enzymatic levels, beyond the speed of light.

A cell does not employ an ordinary generator nor array in morphogenetic clumps by queued formula or contingent instructions. Its instrumentality fuses primal radiations of electrons, uncertainty states of molecules, quantal discharges of organelles, syndications that have no name, no cause and no effect yet furrowlessly weave fabric such that tissues coalesce into creatures inexplicably and outside time.

Chromosomes are articles embodying termless chains of which they are neither the magistrate nor agency. They are marks on a drum, bells in a melody. Perhaps the drum is the skin of time, beating out cacomistle and snail.

Ravenous subcellular mites, while attempting to devour each other, instead fuse into aggregate structures swelling with winsome gametes. They not only invent membranes and sex by their entanglements but spin the fairy tale inside which we find ourselves today, immersed as flesh. Desire *does* have a finite and meticulous beginning.

Modern technology, insofar as it deals only with links that are extant, ignores hundreds of millions of years of microtubule organizing centers and RNA aliases, emerging mitochondria and Golgi bodies, bent micro-rods and looped threadlets, macromolecular amalgamation, subcellular lysis and fusion—system depth and intelligence, system profundity and danger.

The shadow eclipsing the world is fundamentalism—vacuous yet passionate convictions: Islamic fundamentalism proclaiming martial theocracy, Christian fundamentalism launching crusades, Jewish fundamentalism demoting God to real-estate agent, scientific fundamentalism prostrating before a revealed flow of molecules.

Don't be fooled. Science is jihad in another form, issuing its fatwa to subdue the planet and subject it to an idol.

We are in the grips of faux pious factions, each laying claim to a liturgy inscribed in some holy book, each convinced the others are infidels, puddin'-heads, or worse, each willing to excoriate reality, to destroy nature, to make its point.

It should not be long before evangelical Christians, self-proclaimed "born agains," and other conservative cults and white supremacists formally revise their religion, removing Christ from it. The life of the savior is becoming more and more an embarrassment to them. To all appearances he taught humility, forgiveness, peace, renunciation of wealth, and the clarity of personal vision. He drove money changers out of the temple and turned the other cheek; he advised those in glass houses not to cast stones; his method was to find the Kingdom of God within.

None of these are on current evangelical agendas, which are distinguished more by intolerance, self-righteousness, accumulation of goods and power, emotional bullying, institutionalized bigotry, drilled cold-heartedness, and replacement of true visioning by sanctioned revelations. How could anyone think that Christ instructed his disciples to blow up abortion clinics and steal from the poorest people on Earth? How could anyone believe that Christ modeled judging and controlling the behavior of others in lieu of engaging the demons in one's own soul?

He didn't say to build churches and papacies; he didn't found ministries or select pomp, vestments, and passwords. No gospel has him preaching that eros between members of the same sex is abomination. He said plenty about unconditional love, the miracle of faith, and rendering unto Caesar what is Caesar's, but nothing about prayer in the schools, property rights, the sacred prerogative to bear weapons, or national patriotism. He lived the ultimate sacrifice, the crazy wisdom of the Resurrection. His eucharist was alchemical transfiguration and spiritual communion with divinity, not their consecrated reenactment in bread and wine.

In truth, neoconservative Christians have already expunged Christ from their church and launched a new theology without him. They have sent their petition directly to the primitive god of war, ancestor of Yahweh and Zeus, and asked him to anoint and exalt them and to slay their enemies. They no longer need Jesus of Nazareth or Christ the Lord and would no more welcome his return or the appearance of a true messiah than the

Romans did.* We have come full circle to the babe in the manger as an unknown thing, a wonder yet to happen, much-needed good news for a sinful world.

As we adhere to eroding scenery, the eye of attention the Egyptians knew as Horus (hawk) tracks.

Our lives are not things or substances; they are views, access points on creation.

The lynchpin is inserted at an obliquity that cannot be plumbed.

The central thing holding back the human race is a failure to recognize where we are, where this all is taking place. No wonder blind battalions clash by night. We come into being mysteriously, cloned as electrified dolls inside gourds clasped to a stormy spheroid, its alembic of rare and solidified gases spinning centripetally, orbiting a naked star at runaway speed we do not even feel. We inhabit, in fact have arisen from, layers of magnetized mud in this flask.

---

*Is it not ironical that First Lady Laura Bush's favorite piece of fiction is Dostoevsky's "Grand Inquisitor" episode from *The Brothers K?* In the Russian narrative, Christ, returning to a troubled Earth, is taken into custody by the authorities and chided for interfering in religious matters. His Inquisitor explains that the Church functions far better without him. His presence is messing things up, and he should go back to where he came from. The mark of great literature is that it can be read backwards or upsidedown and still retain its conviction and power.

Throughout just about all of humankind's tenure, belief in a divine plan rescued us and inspired beneficent institutions. Our ancestors collaborated in remembrance and continuity. Newton and Darwin still presumed such a blueprint and sought the clews of its maker.

Our presence, indemnified by its own miracle, used to be nature's central event. Now the study of nature has become man's pretext for religious inquiry, his feckless quest for a reason to exist—an investigation that has gotten more and more frantic and desperate by generation.

With the refinement of physics and astronomy into elemental biology, a divine order has been shattered, its particles minced into ever finer nothings. Now, evolving complexes get cleverer only as they accrue synapses to win the war of predation. We are conscious not to ponder but because ganglia run more effective eating machines. Some, in fact, educe that metabolism has nothing to do with it; the animal body is mere encasement for competing genes.

Chemicals go where gravity and heat draw them; they are happier that way. The lynx, like the mouse, like the gypsy moth, seeks comfort and protection outside and inside its bodysuit. Once hunger is assuaged and shelter gainsayed, whims and amusements rule. The corollary is that consciousness is a chemico-cinematic illusion, conducting rootless impulses along trunklines of protoplasm.

Materialism leaves us devoid of any reason to exist, except that we are. And scurry and primp like spiders in cosmic dust....

Science has the tiger by the tail, not the heart, not the brain, not even the throat. The scavenger hunt of the best and brightest for the latch to the Western paradigm rummages through the mortar of our situation, a brilliantly mad civilization tearing apart its own fabric as it dissects cells, atomic nuclei, their particles—the physical forces holding together matter and space—cannibalizing the ship that is carrying us through the darkest of seas. We have become, all of us, ritual cleansers, magisters of superstition, idiot savants, repeating the same mantras over and over. We have no idea what to do with our pathetic one-act, except to convert it to sales, class warfare, recreation—and finally debauchery.

There are no ethics (how can there be ethics in entropy?), so we elevate profit and greed to the only possible motives (plus a scintilla of curiosity about how much worse, more exploited, and mega-corporate, in the absence of rectitude, this can become). We pretend not to have real lives, not to have meanings, so there is no reason to halt the butchery. Ducks and chickens are just protein machines for consumption.

Science as a mesmerizing sermon is driving history, enveloping the archetype of the planet, grasping all objects and species of experience in its gaze.

In our haste not to fail science we sacrifice what is innocent and beautiful everywhere, every butterfly, every iris and toad. We ignore our selves while searching for other things—so-called concrete ones—as though to validate, by deconstructing, the ground on which we stand. Yet that ground *is* our being. Tallying bottomless parameters, tending a crypt of dying stars, we turn away from not only God but our own existence.

We have entered Huxley's *Brave New World* and Orwell's *1984* simultaneously, the worst of all imaginable conjunctions, the terrorists of Frank Herbert's *Dune* massing in deserts and slums: a trifecta once oxymoronic. Billboards soliciting our bodies for neurosurgery clog the foreground with the naked blue pod she will receive in return. Police forces of the Patriot Act patrol what is left of the background.

Where will the children of the next millennium play? Why are their uncles bulldozing clover to make way for universal Wal-Mart?

Matter is all and only energy. That is truistic. What then is mind? Mind is an effect of metabolic action on carbon chemistries deciphered in ganglia to prompt agencies and actions of creatures. Yet, at the same time, mind has no existence in matter or energy. It is off the chart, meaning not that it is above the calibrations of the chart but that it is not measurable at all.

Mind is a molecular, or meta-molecular, force, acutely so in acts of hexing, healing, and battle. Unconsciousness transforms everything else while becoming conscious. The witch doctor and voodoun rule the fog of war. Not commissioned armies, not weaponry as such, and certainly not policy strategized and sent down the chain of command.

If lives were the same length or people knew exactly when they were going to die, existence would lose its edge. The uncertainty of death's moment is the single thing that keeps us real.

If we dwelled here forever, this would not be heaven but hell.

The emptiness of death gives life its fullness. Only love—innocent, unrequitable, meadowlark love—transcends mortality.

After the Indian Ocean earthquake and tsunami of December '04, the Tamil Tigers of Sri Lanka and Islamic rebels of Sumatra called truces and joined forces with their central governments and international workers to provide relief for the homeless and bereaved.

These wars are affectations. Human beings carry them out because they have nothing better to do, nothing else that holds their interest.

They train for battle because nothing more audacious challenges them; they attack in the absence of deeds nobler, more sincere, more real. They are ready for something better.

The ocean gave them something better to do.

In drought-stricken zones across the cosmos, rivers dwindle to ponds, then pools. Packed with all manner of reptiles and the biomass of local river horses and their young, these oases are death traps for the impala-like, baboonoid, and plovery creatures who dare to sip there, to wet their coats and feathers and bathe their cells. Bodies are dragged underwater; troglodytes battle for bloodied viscera and tear them from one another's grasp; skirmishes escalate and spread onshore. A spasm of ego, greed, savagery, and dominion, it is a sorry commentary on the birth of consciousness. But absolutely necessary.

At the end of its life cycle the water hole is a mud puddle from which the last hippopotamus and all except one crocodile flee, each to be fried in the microwave of a star before it can reach new water. The final king crocodile is baked into the hardpan, its bones fossilized, over which, when the rains return, a new river will flow.

Long ago "being" sought itself; now it searches despite itself. Our civilization has fallen like a drunk onto the sidewalk, into the particles composing it, empty, disperse, higgledy-piggledy. We are mind trapped in a dungeon with no keeper, no maker, no guardian.

Even priests, in the sanctuary of their practice, worry that God is an irresponsible truant who has gone to sleep and forgotten us. What they seek in the starry coverlet among pyres of worlds is the flint of their own awareness, to sound the alarm and, if Divinity cannot be awakened, cannot be apprised of the spreading damage, to touch those poor itinerant humans and, in embracing them, reawaken God in themselves.

Take us, who have made us. Reclaim us, who have delivered us here. Restore us, who have scattered and suspended us in dimension. Reclaim us, greater Earth and Sun, so that we are safe again to drink from your chalice.

Ah, but science tells us that there never was a golden time. Not when animalcules gnaw and devour fellow a'cules from planetary dawn, when waterholes are killzones at which wobbly whelps are disemboweled by hardened predators, when soldiers slay civilians Babylon unto Babylon, solar system after solar system, hunters of Ice Ages roast ancient jack rabbit and quail on all worlds. We may pretend that we are more spirit than shade, but we have acted like vampires in this realm, every last ragamuffin of us. Despite fancy haberdashery, we are little more than ghouls, furred and semi-intelligent wizards convening around campfires and factories.

Souls bob up like corks in water. Newborns matriculate as idiots, fuzzy as bean pods, immaculate in ciphers. Toddlers come to their senses, sniff, find bearings, paddle on currents and gradients; defend their survival.

We arrange lives on the false premise that we are. But nothing is.

The dead fall off the edge, beyond retrieval. Everything perishes, vanishes.

We are at the bare beginning of the real journey, in bodies great and tiny, imaginable and unimaginable, through ultimate darkness to ourselves, to an exaltation that will hold life in place until the next conceptionless universe, en route to the fiesta siesta in which all this is redeemed.

Jeffrey Dahmer's most compelling and ominous explanation for seventeen murders, many of them followed by dismemberment and cannibalism, was that, if we are mere blobs of stray muck, robots from the lagoon, what reason is there not to try anything, to taste what there is, to partake of the heterogeneity of carnal appetite, even necrophilia?

What more did Leonard Lake and Charles Ng have to say for themselves, the one boasting laconically of his prolonged tortures of kidnapped women on home-made videos, his lackey penning cartoons on a Canadian prison wall of the body parts of his victims returned to their relatives like prizes in silly sacks? What gave them confidence that their narcissistic drubbings were cute fantasies, that other people weren't as real as themselves? How did Richard Allen Davis bequeath himself Polly Klaas? From where did the proud cackle of Dennis Rader, the

BTK killer, arise as he taunted Wichita for thirty years, this pillar of the community and church elder, strangling women while contemptuously apologizing for his sexual "problem," boasting to the police of his efficiency and ingenuity? What made John Wayne Gacy think he could masquerade forever inside Pogo the Clown?

We have been moving toward rather than away from Mr. Dahmer and Mr. Ng ever since. Bush the Younger's Administration speaks "Dahmerese," applied albeit in a more socially acceptable fashion, with the same horrific results (Abu Ghraib, Guantánamo, Falluja, Tajikistan). If other people are fodder and collateral, then no one is safe from anyone, from any delusion or sadistic inquest, however contrived or inflated.

Joseph Edward Duncan III proclaimed himself a true American, his anti-pedophile adversaries no better than Nazis (per his blog), as he stalked an entire Idaho family in night goggles in preparation for hammering them to death and snatching nine-year-old Shasta Groene and her brother as his trophies. Those folks piously wringing hands and squawking for permanent incarceration of sex offenders don't get it. You can't lock up the human chimera, the rude prodigy that birthed us. It has been traveling inside our race since the voice of consciousness spoke. The strident righteousness of commentators fends off the vile whisperings of their own souls.

What is missed is the desperation, the desperate longing, to locate something here that is as real and powerful as these nihilistic acts seeking recognition, the part that was not nihilistic once. All it would take is a humble turning toward the desperation and asking *it* to lead—giving it space to breathe instead of crimping it by phony guilt, instead of being all over it with dry morals and crocodile tears, then castigating ourselves until we can no longer weep.

It is no use wringing hands and repenting, or hiding from oneself, while continuing covertly to partake.

Until we recognize each other's hopes and dreams, we will stagger to the left hand of darkness, until we accept the ardor in our beings, ugly and cruel as some of its expressions become, as a thing in itself as powerful as the acts—in fact more powerful because the acts are mere zombiisms compelled by empty words in shattered minds.

All it would take, to get us out of this mess, is acknowledgment that we are not so bad and heartless after all.

If the Sun were removed, the Earth would roll outside time, an anchorless cinder with no past or future, displaying heirlooms of once-mighty bipeds, derelict skyscrapers of metropolises crushed by gravitational endgames into cuneiformed stone.

There is truly nowhere to turn in this, or the abyss of unconsciousness, except to an undulating, boundaryless breath—to the equally bottomless chasm out of which, into a primal sea, the jellyfish came.

The great baal, unlike native wind and tide, operates by stealing energy. Man-made machines with their statistical grids run everywhere, without consciousness, without reference, without context, without pity. Bank ledgers flutter like dessicated leaves in November's breezes, keeping score among individuals, groups, nations, establishing and erasing bottomless debt, ephemeral triumph, the raiments of profit and loss. These numbers are everything and nothing, and they can be ground to gobbledy-gook in an instant by an exploding sun or errant pebble, a crude, primordial whack from outside the system. Cessation of all hedge funds.

We have lost touch with the sentinels who maintain space between worlds, who sort one star from another, looking both ways from their watchtowers.

Morning tries to spill out, but not to them. Night tries to bury itself in a foxhole of stars; they track true Night.

The animals are in sullen rebellion. They refuse to talk to us anymore. They won't allow this "king of the jungle, annointed of God" shit. They get it. A snake gets it; a fly equally gets it. None of them will yield a morsel of real knowledge to us (or high moral ground), vivisect them though we will. Fish are mum. Elephants and apes weep inscrutably in their dungeons.

We interrogate and serve summonses. We appropriate and ablate. We act as though we are in total charge here and running our own show. We coerce everyone; we deputize ourselves in realms in which we have no jurisdiction, wherein laws we concoct don't apply. In sham glory, we are running scared, on total empty.

The molecules of the old universe respected two things—gravity and heat. So why do their disciples now rush like famished mites to the festival of biology? How did today's universe get from just thermodynamics and gravity to information (that is, given austere dynamic constraints)? Then how did it convert loosely strewn static into bodies and consciousness? How do rank and rowdy molecules twirl into orderly embryonic disks, the most sophisticated imaginable objects in a physical realm? What, if anything, is added to concatenations of matter, energy, and information to spur their disequilibria toward life?

That is *our* question, science's question, and there isn't any rivaling it. All possible answers lie outside the present reigning scientocracy, though not outside millennial science. The universe is empirical all right, and it is also a vast, self-regulated telekinesis.

Mind is the source of matter: that is the only half-reasonable explanation. The primal state of consciousness from which matter arose would not look like consciousness to us, for we are deep into its density and infinity, far removed from primal lucidity, from pure scintillation. Yet, as sure as gravity lassoing moons, from mind's archetypal seed originated atoms and then, along starkly plowed molecular trajectories, habitable shapes emerged.

Leaving its metaphysical abode, consciousnesss transits through holograms of dust to become the billion synapses of all nervous systems (its expression at a lower octave). Yes, yes, matter is consciousness in another form, even as consciousness is matter sublimated to its primordium, viewing its origin nostalgically and vastly altered—like a gorgeous sunrise—from across the entire universe.

The way that embryos matriculate on all worlds provides a morphological map roughly resembling the itinerary of pure mind through tissue structure to minded consciousness. That is what an embryo is—an exquisite four-dimensional flowing sculpture of the consciousness/matter transformation in the universe. The subtleties and complexities of this unfolding are especially evident on cellular and molecular levels.

Matter reenacts the seminal event of its relationship to consciousness each time a new life form emanates. The personality shoots from some sort of hyperzone into a crude lump in self-organizing biomolecular space. Once design is hexed by the organismal field, the homunculus unravels by atavistic prerequisite, under counter-entropic dynamics, from bacterial precursors into autopoietic hydrozoan and saurian clay, through marsupial and simian molds, to meet the face of an angel in the temple of the mammalian womb. This is divine investiture.

To attempt to assign physical/spiritual development merely to the intersection of heat, mass, information, peregrinations, and the statistics of natural selection is to miss a profound chaos-phase, kinetic/telekinetic force pouring shape and energy into starfish and caterpillars everywhere.

Letter from Joe Richardson, October 9, 2004, Holbrook, Arizona

"Two days ago my cat watched as I changed my oil in the car. Cats have a natural curiosity about them. She looked at me as if I must be the weirdest critter she'd ever seen. I tried to tell her that sometimes humans do things that are at least dumb (if not senseless) like building an entire civilization based upon the internal combustion engine, knowing full well that some day we'll run out of resources necessary to maintain that contrived civilization, but she just laid there and yawned....

"The bugs, flies, and flying insects are all but gone for this year. Outside, it is getting colder. In other parts of America the leaves are falling from the trees. The same may be said for the world and the age we are ending."

The heart of the universe settled long ago (and is still settling). A massive turnip anchored to the bottom of creation, it splits into worlds because turnip-hood is gravid.

A temporal sun rules the sky, coupling the Solar System to the Milky Way. Fire is a hydrogen phenomenon, transmuting essence, releasing energy.

The Sun is not just a sun, a chemical, thermonuclear furnace; at its nucleus rests a finer, more cardinal, beyond-the-speed-of-light light.

All animals are slow-burning fires—gopher, parrot, and clam—scurrying, vibrating, siphoning. Embers smolder inside wet suits of ray and eel. A plant—blue, yellow, orange; red, violet—is a droplet of star.

What if the Sun were understood as pure illuminated intelligence blistering into this dimension? Then that intelligence would radiate equally pure hydrogen atoms, igniting with terrific energy from a transmuting core into massive solar flares, breath disseminating into space-time. Stars are self-knowing sources, self-lighting fountains, providing bodies of worlds, helices of DNA, other plans.

Consciousness *is* the origin of matter and not some trick of sums of random stuff adhering into tissues and neural fields. As such, it explodes from the surfaces of stars into figurements in thermal springs, atomic draff coagulating into mind again, baptizing each creature inside the temple of life.

Galactic embers have consciousness imbedded in them because mind is already intrinsic in every particle of matter, stippled onto even water and air.

Each sun is a welt on the relativistic skin of the universe, a primal gateway to the heart of creation.

Why Weapons of Mass Deception Are More Powerful than Weapons of Mass Destruction

Mind is more deeply rooted in nature than matter is. As the wellspring of matter, it cannot be razed or immolated by material forces, however truculent and massive. As Tsoknyi Rinpoche told his students in Kathmandu (well before the Maoist insurrection): "You may want to smash a painful emotion to bits, but you can't blow it up with a nuclear bomb. Even hundreds of thousands of nuclear bombs detonated at the same time will not stop dualistic mind from creating more emotions. If someone were to kill every human being in this world, dualistic mind would still continue making emotions. Through the power of karma, all these minds would take rebirth in some other world and continue in the same way as before."*

You can't get more basic than this. Darwin and Newton are "bad," Marx was bad in his own way, but Buddhism is the baddest working statement that the local animal brain has concocted.

Although bodies are summarily crunched, although plan-

---

*Tsoknyi Rinpoche, *Fearless Simplicity: The Dzogchen Way of Living Freely in a Complex World* (Kathmandu, Nepal: Rangjung Yeshe Publications, 2003), pp. 46–47.

ets and whole universes can be obviated in a gulp, creature mind is unabashed. It travels through even the heat and pressure of stars, which is how worlds slagged from solar rust, just a few million years after cooling, ineluctably populate with sentient forms.

Mind may be dualistic and chimerical, but it is fundamental; it penetrates the basis of organization, even its own. Its ornaments and plazas keep coming, its thunderstorms and melodramas, even after Hiroshima, after Bhopal and Rwanda, even after cremation, after fifty bullets have been pumped circumstantially into its corpse.

An astronaut accelerated by rocketry finds himself suddenly beyond night and day, in a zone where sunset and sunrise are mere perspectives about a wick.

But true Night-and-Day described by Hesiod in *Theogeny* cannot be maneuvered outside of, cannot be exteriorized from any vantage. It exists by inherence.

What we never get to see are the skeins of yarns, nocturnal and diurnal, that have been wound around themselves billions of times to plug us into a Sky.

In a dark room a candle glistens, tossing shadows onto walls. In a dark universe, mass-condensing suns and galaxies illuminate time and space. At the core of mind, something like a candle or bonfire ignites. Self-emblazoning, intrinsic, it cannot be brightened from outside.

Turn inward to the source of the glow and behold mind's stark essence. Rigpa is at the heart of every "guy" or "girl," every gourmet meal, every seduction, every crime, every rap, every fury, every grief. Rigpa is empty, passionate; steadfast, elusive, inescapable. It is the source of "Fuck you!" and "Love ya" equally. When you grasp at the most joyous, timeless epiphany—a child, a meadow, a song (*"Dream lover, until then,"* or not)—rigpa is its substance. The circumstances are all illusions. When you feel wrenching melancholy and anguish— devastation, terror, betrayal, separation—these too are empty, sustained by the same candle within. Find the candle, the lamas say, and you find your way out of the maze of samsara.

But you can't just meditate on mind and not live because life's drama and panoply mark this place in the universe; you also can't *not* meditate on mind without forsaking the sun at your center.

Across the immeasurable infinitesimal gradient from grief to ecstasy, from puppy love to abandonment, is not only everything but nothing, a paradox so diverse and wondrous, so brutal and subtle, that its application suspends us in myriad events with no solution, no meaning, no closure, composed of meaning itself.

*A Polemic**
Jihadist attacks, especially the more gruesome demolitions, incinerations, beheadings, as well as biochemical and nuclear

---

*The polemics in this book began as parts of a single essay. The only one that does not appear in these pages, "Abu Ghraib: A Howl," is included in an anthology, *Abu Ghraib: The Politics of Torture* (Berkeley, California: North Atlantic Books, 2004), pp. 123–140.

threats, while publicly decried, are covertly welcomed by Likkud-
ists and neocons, as they provide a worthy menace against which
to mobilize armies, a barbaric enemy for their propaganda
machines, an irrefutable provocation by which to sway elections, a ruse under which to invade and occupy sovereignties.

America's Boy George seems positively thrilled to have an
opponent committing vile, unapologetically "evil" acts—vain-
glorious as a frat boy at the stunts of a rival house. Jihadist
blatancies spare him etiquette and compromises of diplomacy,
the feigned gesticulations of a clemency he does not feel. Since
he cannot be a samurai or statesman, the role left to this ambi-
tious self-proclaimed born-again is "moralistic thug."

How far is this guy anyway—an instrument of casual violence, a scion of unearned wealth—from fraternity cut-up,
from pledge-master of the Skull & Bones?

In his gut he wants to curtail government charities, poverty
programs, educational and after-school services, health-care,
social welfare, environmental protections—as much of the
safety net and national park system as he can get away with
dismantling—because it is taking wealth out of the capitalist
game for those who are winning and having fun, whether cheat-
ing or not, and that's America to him (and his crew). Endless
war is a perfect excuse for emptying the treasury and then
pleading a paucity of funds for anything else—sorry (but not
really).

Piety and gamesmanship are the poles of W.'s paradigm,
sanctitude and prerogative. He loves being able to declare,
"Bring it on!"—three words that (despite his later disingenuous regret) bracket his sum response to the poverty and anti-
globalism at the heart of "terror" and Islamo-fascism—and
then cavalierly show off his golf stroke. In the accolade of Dick

Cheney: "He's got the biggest balls of anyone I've ever seen!"

The terrorists' lack of a regular army or "bricks and mortar" factories makes them a potential "bum of the month" candidate for a politician trying to build his resumé. After all, if someone smacks you, pick out the weakest nearby chump and flatten him: "That shows what you get when you mess with me!" W. relishes the opportunity to posture self-importantly at others' expense before smashing and humiliating them with superior toys. He believes he is not only winning but kicking ass because we have the single military-industrial juggernaut and God is on our side.

Yet the asymmetry of the battlefield, far from guaranteeing him victory, works against him, favors him superficially and temporarily. Any munitions made on this planet can be used by anyone, rich or poor, smart or dumb, who can get ahold of the parts and operating manual. There is no exclusive on technology. With each new weaponized invention and proliferation, we are subsidizing our own undoing, for we are creating raw, anarchic power and releasing it in the global bazaar.

The master of rope-a-dope when it comes to domestic electioneering, W. is a gull of the international variety, as he rashly expends capital, overextends state-of-the-art militias, consumes ever more nonrenewable energy sources, and drains his nation's wealth for generations to come, perhaps all generations, into welcoming Asian and Arabian banks, in an attempt to obliterate agitators waving red capes and prayerbooks at him, trying to lure him into phantom battle, sacrificing their bodies for the real armies of the future. Bush is thrashing at an opponent who *wants* him to "smart-bomb" to his heart's delight and pull every fusillade out of his bag of tricks, to think he is swatting flies. When he exhausts what he has, along with his bravado, the cobra calmly will strike back, either now or fifty years from now, whenever it is ready.

Yes, for the thousand thousandth time *ad nauseam,* the world is better off without Saddam. He was a vulgar, conscienceless butcher, bellicose, narcissistic, criminally psychopathic. The Baath cabal was routinely corrupt, capriciously abusive and murderous. Iraqis have every reason to relish their new freedom. But this is hardly the main event. How many died, were maimed, and lost their families in the snuff ritual? How much damage did Bush and company inflict in bringing down the regime so as to lay out Qusay and Uday's corpses and pull Saddam from a hole to check his teeth like a goat? Who exactly are the barbarians?

The word "freedom" has become W.'s nostrum, a carny's motto for what will make you happy—a buzz word for capitalist and peasant alike, serf and bedouin. His "freedom" means incorporation into the West and the global economy; it means having your ancient forests stripped, your oil pumped into 7-11s; it means toxins flowing unchecked into your rivers and veins; it means losing your traditions or having them bowdlerized into quaint commodities. It means making your children slaves of fashion and markets, or just slaves. It means giving up your gods and songs, having them supplanted by corporate forgeries.

Anyway we can't "free" everyone; in fact we can't free anyone. As the young Dalai Lama legendarily told the Chinese general sent to convert him to the Maoist cause, "You cannot liberate me. I can only liberate myself."

Reckless, flush with hubris and jingoism, the neocons don't get the "bigger picture." They don't care if innocents and civilians are slaughtered or mistakenly imprisoned and interrogated along the way or if they are inflaming a generation of anti-American warriors from Central China to West Africa. They conflate class privilege with divine mandate. The impoverished villages of the world fall outside their purview. They don't see families cowering before Humvees full of First World soldiers aiming automatic weapons at the Resistance. They don't get to gather up what is left of a child's body after an air strike has done its work. They don't know what it feels like to be dismissed in your own land as human refuse by self-important proconsuls, redneck mercenaries, and dumb chicks of the empire. As for their ballyhooed preemptive espionage, they are mostly fools and bullies, a lethal combination: "they torture the wrong people ... they torture too many ... and for too long. It is useless to ask everyone everything you want to know."*

War is a crude, unimaginative, and heartless instrument, even for liberation, even for overthrowing homicidal tyrants. Those who least understand a situation make the largest, most awkward moves, as if everyone before them had missed the point. Only cocky amateurs who mucker into power think they can accomplish anything. But blatancies conceal enormous depths of subtlety that cannot be crushed out of existence. Blundering warmongers and do-gooders unleash an unimaginably inextricable skein of counterforces that go on rebounding and uncoiling into the future. Do you have any idea how much depleted and non-depleted uranium has been dumped

---

*G. Jan Ligthart, "Letter to the Editor," *The New York Review of Books,* February 10, 2005, p. 43.

on Afghanistan and Iraq in the name of freedom? According to some reports, the equivalent of 250,000 Nagasaki bombs fell on Iraq alone through 2004, four million pounds from artillery shells and tank armor in 2003. Vaporized nanobits of this stuff are now blowing all across the planet.

One can accuse Saddam and his associates of gassing Kurdish villages and slaughtering Shi'ite civilians and indict the Taliban for rabid brutality and repression, but America has brought its own, far more profound and lasting apocalypse to these countries, one with a half-life of billions of years measured in tumors, extinction of native plants and animals, breaking of lineages, and exotic birth defects. This inconvenient by-product of "emancipation" is all but ignored or discounted by the press as if inconsequential.

Meanwhile Brother Bush is playing the big game, having a good time: "Look at me. What a ride!" Like someone dreaming he is President, he expects any moment to wake up.

To a troubled citizenry and world, this is not a joke; this is not a dream.

In an unguarded moment while still governor of Texas, his expression and intonation feigning female desperation, the president-to-be mocked Karla Faye Tucker whose appeal for mercy he had just denied (in part, to punish her for consorting with Bianca Jagger and Larry King). "Please, please don't kill me," he said in cruel mimic of her voice.* After her execution he impersonated sympathy and prayed aloud for her soul like some ham preacher.

---

*Sister Helen Prejean, "Death in Texas," *The New York Review of Books,* January 13, 2005, p. 5.

W. may camouflage his smirk at one level; at another he cannot. Nor can he disguise, by patriotic poses and churchly sermonizing, the violence in his heart. This commander-in-chief, though never in battle himself, let out a crazed whoop upon hearing that the "Shock and Awe" bombardment of Iraq had begun. What legitimate warrior would so demean the battle? Boy George has a kind of primitive savagery, even a raw sadism in him that passes as mere cowboy, a childlike tyrannizing that has carried into his adult persona unchallenged, making him more a demagogue/warlord than a compassionate conservative, less a missionary than an anti-Christ* (though "lover of freedom and democracy" is his plenipotentiary alias). He does not take war seriously, not in the way that it is serious to those who must endure it firsthand, who experience the demolishing of their homes and massacre of their loved ones. He knows nothing of the gods who still rule the affairs of men and women from Olympus.

---

*I know this sounds like exaggeration or petulant vilification, but how else would the anti-Christ appear in our times? He would dress in stylish suits and behave like the most upstanding citizen. He would pray publicly with seeming earnestness and then smartly salute the American flag. He would claim to be a true Christian, a stalwart defender of the faith. He would utterly fool those who lack the true Christ in their hearts.

Don't expect an anti-Christ with horns and 666 tattooed on his forehead. That's just tabloid Rapturemania. Don't expect to see him engorging himself with blood and entrails. But if you attend closely, you will see that everywhere he lays his hands, blood spouts and nature is laid waste; life is poisoned and turned to ashes. While his minions slaughter at a distance, he can barely suppress his smile.

It would surely be a more legible realm if the devil wore a uniform and placard around his neck, if good and evil were spelled out so that we could tell the righteous from lackeys of Satan. Yet figuring that out *in medias res* is the whole game here.

He believes "terror" is a test of his machismo and resolve so, instead of cultivating wisdom and right action, he demonstrates how smart he can look landing on an aircraft carrier, how coolly he can blow off rules for interrogating prisoners of war and spurn international opinion, how jauntily he can strut to a podium, rakishly wear his hat, toss out retorts, and dissemble with the merest cloud of a sneer on a poker face. He calls it "Texas walking," but, as Senator Byrd nailed it, for all his *cojónes* W. Bush is "all hat and no cattle." Or as a Gaza factory worker told an American patronizing him with shibboleths of democracy, "Don't tell me about my responsibility to vote. You elected a moron."

Adversaries in the Holy Land are trapped in a vicious cycle, as fanatics on either side reincarnate endlessly as each other. The same enlisted men and women—troops and civilians, jihadists and settlers—clash lifetime after lifetime, armistice after armistice, against themselves, while the magnificent reconciliation in their center refuses to be born.

Palestinian urchins at play, struck by a stray missile, awaken in Jewish synagogues; old rabbis crawl as babes on the dirt of Bethlehem huts. An Israeli soldier firing at Palestinian teenagers, ambushed by a grenade-thrower, gasps next at breath, memoryless in Hebron. One-time Hamas suicide bombers stride as privileged tots down avenues of Tel Aviv; their victims hail rocks onto Star of David tanks. Officials condemning houses in Gaza are a few blinks from collecting playing cards of Palestinian martyrs. Only the slim, unreadable thread of karma connects one life to another, one playbook to another.

That is the way energy moves on planets in this zone. Violence is blind because it does not see that the only thing it can strike is its own projection. The only thing it can harm is itself, distorted through guilt, ambivalence, and loyalty. The only thing it can become, by physics alone, is what it girds itself against.

Maybe another hundred thousand years of this, presuming we educated apes do not exterminate this sphere, will make the lesson clear.

The Buddha said, "What you are is what you have been; what you will be is what you do now," and Padmasambhava legendarily restated it: "If you want to know your past lives, look to your present conditions. If you want to know your future life, look at your present actions."

This parable yields to no moral forecast, no simple formula; the future cannot be assured or even measured by deeds, the immediate results of those deeds, the good or bad intentions behind them, even the feelings supporting those intentions, nor by any sublimation of any combination of these.

We are beyond elaboration and cannot be guaranteed.

Invest only in loss. You cannot plan for fortunate rebirths or fair rewards; you cannot indemnify a comely body or even a human one; you could become a slug or horsefly. You cannot collateralize *anything*. You cannot even play the game as if you knew its rules. The road to heaven is a boulevard of land mines the moment "heaven" is its name.

There is another judgment and it is so basic and equitable it will turn prayers into curses, nightmares into euphoria, theft into charity, bounty into debt. By its algebra, despair will become celebration; blunder, revelation—and *vice versa*. Prior to language, prior to embodiment, the universe is more literal and final than we can imagine. Its tortuous star fields and molecules are an absurdly precise rendition of karma.

When someone foresees an event, either symbolically or telepathically, or, in the aftermath of an event is presumed to have precognized it, she is actually creating that event, though not in the manner we might suppose.

The seeming meaninglessness of life is resolved by cumulative existence. The effort to raise the sun each day is the crux of reality. This is a team game, with no opponent.

Stray Commentary on a Lunch Line
"Buddha was a Hindu just like Jesus was a Jew. Same vibration, different frequency."

We are historyless now, since the atomic bomb. How can history proceed when a single act of a fanatic or suicide bomber can obliterate it? How can ordinary business be conducted when smuggled uranium and detonators will annihilate hundreds of thousands in a snuff, their shops and living quarters, the future of their planet? How much longer can the imperiled city delude its populace?

Nothing means what it used to; no one can earn the gold watch. The world has been hibernating in an unmarked bomb shelter for six decades, pretending not to notice its own disappearance, passing across a hypothetical century line, its gate at 9/11/01. The young are now so bored, so restless, display such bogus invulnerability.

We are secreting something that refuses to move forward, that eats time without constructing a web.

Computers are not our saviors; they are hooded hawks that magnify the depth of the maze in which we wind compassless. They log a poor fraction of the integral universe, keeping us occupied and irrelevant.

Binary (0/1) is not the only form of information; there is also a physics of empty mind, a mathematics of love.

"When a new gift, a new possibility, is given to the earth, it is always presented in two ways," Rodney Collin wrote in 1955, "in unconscious form, and in conscious form. In the hydrogen bomb we recognize the unconscious form of a power hitherto unknown on earth. We await the demonstration of the same power in conscious form, that is, incarnate in living beings."*

The real perils are privilege, attachment, "tantrumitis," maudlin glory, and the projection of battle onto imaginary gods.

---

*Rodney Collin, *The Theory of Conscious Harmony* (London: Robinson & Watkins Books, Ltd., 1958), p. 191.

Our risk in getting born is that creation is cruel and sadistic or, even worse, banal and antipathetic. Since we are well-intentioned, agreeable people, how did we land here among the hooligans and predators?

Looking always *there,* at the skirmishes of nature—at novas imploding, galaxies colliding, sharks ravaging, particles anni-hilating, mercenaries under orders—we see outside not inside the event. If we are in the universe, *what the universe is* is in us. If the universe is diabolic and malign, our innocence is affec-tation. If the universe is random, we are flukes. If the universe is a machine of predation, we are incapable of even a single moment of compassion.

To rescue the universe requires but one empathic act of a sentient being on a planet somewhere among the galaxies. Once that happens, any creature can touch the invisible rag and be transformed into a bodhisattva. Our essential goodness (Anne Frank acknowledged even as the Nazis closed their cordon around her), our capacity for cosmic laughter, the playfulness of simple animals, their guileless love, their nurturing of sprat and foals—these are axiomatic, the last place the little metal ball drops, after bouncing every which way on the Great Roulette Wheel.

What is amazing is the industriousness and sanctimony with which humans labor to deny this. Nowadays admission of altru-ism is a sin; the world is all business, nothing personal. Funda-mentalist religions impose a good/evil dichotomy whereby an adherent tries to stay on the side of "good," in the company of good, against its other. This suggests "here" and "somewhere else," good intentions and those who thwart them, righteous souls and nefarious punks, soldiers of God and abject terror-ists, infidel defilers and austere jihadists.

The ambition to establish oneself as "good" is a specious attempt to hide from one's own shadow, thus not to be able to transmute it, never to understand that it is not really evil. The central act that brings evil into the world is flight from the Other.

If the universe were wicked, there would not be so many psalms and mantras, pixies and weevils to appreciate its arbors and ponds; such layers of joy and beauty, so much allure; so many round and belly dances, endless to no end.

Any way you cut it, we are headed in the same direction for our absolute nature.

The summoning diphthong of the mourning dove—an amped, differentiated coo—is followed by three confirmation notes. Piping of frogs, fiddles of crickets, hoots of owls, barks of coyotes, clicks of dolphins, Senagalese drums have check digits too. In fact, nature is awash with parity bites, roars, and redundancies, as DNA, skimmed through neural software, organizes into repeating series. Progression is inevitable, aria and howl ordained, no matter the debris and tohubohu in which it is cached.

In pods upon pods, breaching and submarining, ancient leviathans congregate upon summons by South American hummingbird calls, slowed down on tape to "subsonic twinklings," then piped into the sea through underwater speakers by a former Vietnam War paratrooper seeking his lost soul.* After the whales'

---

*Martín Prechtel, *Disobedience of the Daughter of the Sun: A Mayan Tale of Ecstasy, Time, and Finding One's True Form* (Berkeley, California: North Atlantic Books, 2005), pp. 126–28.

courteous "chattering, hooting, and cooing" responses to this trans-species broadcast are recorded and speeded up, ornate hummingbird phrases fill the room. If these supersonic messages out of the ocean were piped into a garden, hummingbirds would thrill ecstatically to the recovery of their own lost songs.

But there is a whole other event—an aggregate buzzing, chirping, and crooning in hot, sweet grasses—beyond codification, beyond memory, beyond either redundancy or seriality, beyond meaning or no meaning—scrumptious, wanton, bottomless.

Martín Prechtel Retells a Tzutujil Mayan Myth

Our world, which is coming apart and being remade even as you stand here, is in need of many things to sustain it. Some of them are obvious (food for its creatures, air for them to breathe, clean water to swim in); some are not so evident. It is mainly great stories that support and revive the world from moment to moment, the eloquence of extraordinary metaphors that rouse our hearts and carry our spirits into the realm where everything again makes sense and everything is yet possible because everything both is and is not what it seems and the daring, impossible connections of everything to everything else are revealed as both ineffable and ordinary.

Poorly told, such stories dry up like jellyfish in hot sand. Well told, they undulate in exquisite, meticulous, untouchable beauty. Well told, such stories exhilarate, refresh, heal, turn sullen-ness and deadened hearts into glorious, bottomless grief and grief into mournful ecstasy; this is the epiphany of creation.

"Disobedience of the Daughter of the Sun" is one of the

great stories of the Milky Way Galaxy and is told in one form or another among the tribes of worlds across the Galaxy, all of whom understand it in their own ways, re-tell it in new forms in their own languages, and recreate the archetypal rivalries and hubris and beauty of Sun and Moon (in their local forms) in their necessary struggle to bring primordial night and day to their skies.

Myths are zodiacs, codas of a shifting universe. The tale of the tragic romance between the beauteous, misbehaving daughter of the Sun and Moon and tiny, resourceful Hummingbird, son of the Ocean, is not only how nature everywhere with its winds and rains and rhythms and cycles came to be, not only how boys and girls find their manhood and womanhood and learn to court each other, how birds sing and vines ravel, but a ceaselessly recalibrating prophecy of the condition of the Earth, so that it now speaks to petroleum civilization and global warming.

Martín Prechtel is one of the elite story-tellers of the Milky Way, keeping this legend alive, keeping its jellyfish suspended in rainbow-iridescent billows, helping to hold our imperiled civilization together.

Bill Maher's one-sentence explanation for male enforcement of the burqa in fundamentalist Muslim cultures:

"If I can't have it, throw a tent over it."

But Persian ayatollahs say that concealment makes for a healthier and happier society than *Playboy* centerfolds and mini-skirted streetwalkers.

Either way, nudity is intimidation.

In our flying room, wind raps at the outer aluminized shell as we bump through clouds. The oils of civilization ignite at 520 miles per hour. Untold numbers of wrathful deities toss and claw at the membrane, roar to get in. We vibrate, drop precipitously, but, with fires hot, flat long limbs of avian ancestry balanced on an ocean of gases, we rise through cloudy zones, settle atop, and see a mountainous manifestation beneath. Only the sound of Tibetan monks chanting on my CD player into earphones (seat 22A) can balance this terrifying apparition 28,000 feet above the ground.*

What Is This World For?

1. To get to experience the layers and possible layers of bodies, shapes, and myriad waves and currents that flow through them, resulting in subtle balances and sweetnesses as butterflies, lemon blossoms, fungi, winged migration, watery depth, modes of song, and the unmaking/remaking of the great Tree of Creation every moment from filaments and roots out to liberated seeds, through possible generations and rebirths from amnesia into other evolving forms consuming each other's shapes and identities across the vast universal metamorphosis.

2. To recognize, engage, and transcend ancestral, genetic programming, savage instincts, and personality stereotypes

---

* It is quite a thing, to be zipped shut in an aluminum can and have it set on fire and shot into the air. That is how the autistic part of me views airline travel—a premise both exhilarating and terrifying.

that lock creatures into their predecessors' actions and separate them from their apotheosis, core nature, pivot points, and innate tenderness.

3. To bring many out of one and return one and all, altered, to the zero egg.

In the universe in which we find ourselves, pretty much everything is far. The moons of Jupiter, so close that they have names, are many years distant by present aeronautics. The remotest galaxies are billions of eons from Broadway by any conceivable engineering. Even at the speed of light—and you know how far how fast light travels, 186,000 *millas* every second—it would take eternity to reach them.

The trouble with light as a measure for "getting there," like the trouble with distance as a measure for "there," is that, after all is said and done, there is no real time or space in consciousness, no scale in imagination, no separation in love.

In the future, maybe fifty years from now, a giant post-Hubble lens, having been transported to beyond the dust and murk of the asteroid belt, grinds out an image of a blue and white marble ("in bigness as a star ..."*), a previously unknown world circling Alpha Centauri. It is a different "Earth," the first we have ever seen. Its jigsaw continents amid Oceanus are a solar system away, tantalizingly close, yet beyond our reach.

---

*John Milton, *Paradise Lost,* 1674.

To get visitors from here to there will require a journey of three and a half light years, well beyond the mortality of any human crew. In fact, the trip will take anywhere from twenty to a thousand generations of men, women, and their children migrating aboard a village-size spaceship, the precise number depending on thrust, length of generations, and route. Yet, with Earth's ecology in decline, species becoming extinct, and irreplaceable resources running out, plus the mystery and siren of "another Earth," the odyssey cannot be shirked.

An absurd skein of fortuities, mechanical and human, must go right for the mission to succeed—preposterous given the universality of Murphy's Law. The spaceship must be launched in pieces and assembled in space. Once accelerated toward Centauri, it has to function for a period of time perhaps longer than the whole of human history to date—a tall order for the manufacturers.

Its blast-off from orbit will be dramatic and spooky—a massive object ejecting compact but intense flames, moving at increasing speed motionlessly into the black, already outside time, at once modern and archaeozoic. Its after-image is compacted with dust and space barnacles, as those inside history picture its arrival at an inconceivable destination as far into the future as the emergence in Africa of *Homo sapiens* is in the past, long after all participants and their imaginable descendants are dead.

The knowledge of how to operate the ship's technology will have to be passed from generation to generation aboard a habitat hurtling at meters per second through pure space, a biosphere that will serve as midwife, academy, and cemetery for its tribe. There can be no mutinies or plagues, nor mishaps with cosmic debris along the way. *Sans* sunlight between stars, the passengers will have to feed themselves nonphotosynthetically. In fact, they will lack solar energy and fuel for epochs, as they traverse millions upon millions of furlongs of a wastefully extravagant abyss.

Who would undertake such a pilgrimage for the rest of their lives ... and consign their children and children's children to transit through the interstellar wind? (Of course, there will always be adventurers for the sake of adventure, compulsive risk-takers, paladins who honor ideals beyond their own lives— and they will have children who will inherit their choice.)

What will lifetimes be like for humans born and raised anywhere along the journey, never to set foot on land or touch flowing water, too many miles (and millennia) from the beginning to have any conception of Earth, too far from Centauri for it to be anything more than a fairy tale?

The impulse behind the adventure will have to be cultivated and renurtured among children who know nothing but the ship. Without a vote in the matter, they will have forfeited much of their biological heritage. Subject to zero population growth, they must feed off the mulch of their parents and even their own children. For their great-great-great grandparents, anything from before the ship was already a rumor of a hypothetical culture.

At first, video reports from on-board will be a popular Reality show, drawing top ratings for years. Gradually interest will wane. Both actors and viewers will grow old and bored and die. New characters will replace them. Temporal distortion between the craft and Earth will increase. Events on board will decelerate as if by reverse time-lapse; inexplicable anachronisms will occur with ruptures of sequence. From the standpoint of the crew, more will be packed into less and less time back on the home planet, as Earth snaps its synchronism with the ship and speeds on ahead into the future, extending the gap that scissors apart the two worlds, one made of the other. How will communications officers and interpreters negotiate the burgeoning warp?

Plus, years and then decades (and then generations) will be consumed in logging on to exchange a single greeting. How can operators sustain meaningful dialogue, especially once a sender cannot live long enough to hear the reply from someone not yet born?* Will post-modern techies of coming centuries appreciate the expedition's rationale? Will they eventually develop rockets swift enough to overtake the vehicle and bring its grandchildren—willing or unwilling—home?

Can the radio connection be made so sacrosanct that its umbilicus is preserved despite everything else happening aboard both craft for hundreds of thousands of years—planet and its lifeboat? Will the crew's land-bound cousins survive wars, environmental degradation, rising sea levels, their own cosmic collisions? Will humanity become extinct, except aboard the ship? Will the travelers understand the world from which they originated or why they have been cast loose?

---

*In a children's science-fiction novel Robert Heinlein solved the problem by telepathy, which is instantaneous, no matter how far apart sender and receiver are. His telepaths were twin brothers. As I remember it (from my reading some fifty years ago), the one aboard the rapidly accelerating vessel en route to Centauri experiences the strangeness of relativistic time dilation firsthand, as his brother grows old and dies, to be replaced by his son, and then by *his* sons and daughters as telecommunicants, until, far from Earth and still young, he is transmitting to his great-great-great-great grand-daughter, a spirited girl eager to meet her ancestor. Time on Earth seemed to have speeded up absurdly, carrying all in its tide. This raises the issue of what time really is.

At best, unborn generations will be trying to communicate with each other in an ever-increasing lapse of meaning and language. In time, speech will morph into new dialects, then new languages. Words will change meanings, garble, disappear. Subsequent communiqués will resemble a message from *Beowulf,* then vanish into their own proto-Indo-European roots.

Ursula Le Guin describes a religious movement aboard such a fictional migrating city. In her version a segment of the population comes to venerate the sanctuary of space itself—a faith that nourishes the lives of those in transit and keeps them from getting vapid or desperate. On arrival millennia later, many of the passengers do not want to leave the ship, even to check out the Centaurian world. Their fear is greater than their curiosity. The story turns on the schism between those who remain aboard the sanctified vehicle and those who brave the descent onto a pagan planet.*

How is this different from our present dilemma, rolling blindly across history, from nowhere to nowhere, unable to contact our ancestors? We could use a religion like that of Le Guin's pioneers, traversing as we are, beyond time, an unknown sector of the Galaxy, supported by faulty technology, with no future and no origin.

When the ship arrives safely (presuming an unlikely string of serendipities), hundreds of thousands of years having passed, who will be on board? Will it be human? Will we have succeeded in sending our kind across a sliver of the galaxy? Or will someone else arrive, a people already no longer us?

---

*Ursula K. Le Guin, *The Birthday of the World and Other Stories* (New York: Harper Collins, 2002).

Will the space nomads understand how to debark their habitat? Will they be able to access orbital and reentry files from sarcophagi older than Egypt in their ship's bowels? Can they assemble, launch, and land the musty shuttle, clunk, on hard rock? How will they adjust to the exodus from the cosmos onto a tumultuous, complicated world, with wind and hard rain; oceans, volcanoes, and dust storms?

If Centauri's planet is populated by simple primates or primitive vertebrates, will it be ethical for the migrants to colonize it and fructify, substituting their DNA for its own biological destiny? Ethical or not, what other option would they have? Can our metabolism run in another biosphere—oceanic, cloud-laden, yes; mild, photosynthetic, yet accoutered with non-DNA creatures? What if our children can't digest the native "plants"?

If the world is inhabited by bands of Stone Age peoples, is it similarly justifiable to set up camp and take over—or even not take over? Will our descendants bring discord, battle, and slavery? Would such things have transpired anyway among the indigenes?

What if Centauri's planet is inhabited by a far more advanced race? What will happen to our travelers if, after such an epochal journey, they arrive as yokels in a jalopy? How would we welcome equivalent refugees from Centauri?

Why does it seem as though this has already happened?

In a sudden, lightning-like burst, a concussion of the eternal and the temporal, the ineffable and the mundane, ecstatically the soul seizes the emerging coil of the notochord, burrows its proboscis into a new atavistic creature. This embodies its singlemost unfailing impulse: to claw back into the world. Aleister Crowley: "Every man and every woman is a star."*

The act of passing within creation is as soft as dawn through the finest dust suspended in eternity, yet as abrupt as a crest of lightning in a vault. The bioelectrical impulse merges with the tendril of the CNS in a spreading cumulus of cells. Heart and breath co-ignite. Luminosity hardens, becoming syrupy, then colloidal, then neurovisceral. Each embryo puts on the ancestral garments its karma dictates.

Reincarnation yet again.

What is "self"-aware is, to itself, real. Why should life be any less a wondrous occurrence to an animal than a person, should sea urchins, wolves, and wasps savor their existences any less?

Apologists for animal servitude cite an absence of self-consciousness as indication of "nonreality." Moles and snakes, even though from our identical DNA-cobbled carbon-nitrogen base, are considered mechanisms. Guinea pigs and geese are automatons, cycles of electrochemical sparks on physico-instinctual triggers. They don't really feel because they don't know they feel, hence don't really exist.

We alone are self-annointed as real.

Guess what, dummy? It is all real, conscious at the same frequency, projected into phenomena by the same enzymatic circuitry. Other creatures have biographies as poignant as ours. A chicken raised in a factory "farm," never vacating its cage,

---

*Aleister Crowley, *Magick in Theory and Practice* (New York: Castle Books, ca. 1950, no date given), p. xiv.

its eggs stolen, unable to gather a nest from dust, terrified 24/7, drowned in boiling water (to remove its feathers) at barely a month, has a profound biography. A pig imprisoned in a cubicle about the size of his body, then led, aware of his looming fate, to rotating guillotines, contributes his account of the Earth to the universe.

Look at two dragonflies mating, how they clasp together wildly in the summer air, splash ecstatically in the pond. Look at caterpillars hanging from threads in segmented, furry capsules or purling their slinky tractors along the ground. See flies darting to and fro, buzzing atop decaying carbon piles. They are happy. This is what "happy" is.

Bugs on all worlds arise from a single hearth. To squash any of them is to quell a circuit of the great dance. To slice open fellow creatures with blades, sequentially on an assembly line, is to blaspheme heaven itself.

Plus, discarding animal lives ravages the meaning of our own. We cheapen and make ourselves insincere.

You have to wonder what we think we are doing and what solace or verity we are trying to establish beneath the pretense of our lifestyle. We are making this an unsafe universe. For everyone!

In 1970 William Kotzwinkle wrote *Doctor Rat* about a renegade rodent who escapes his scientist captors and hides out in the experimental lab in which he is a subject.* Driven mad by his plight as well as the grievous injuries he has already received,

---

*William Kotzwinkle, *Doctor Rat* (New York: Alfred A. Knopf, 1971).

he comes to identify with his tormenters. To his thinking, the other lab animals are ingrates, as they complain about experiments in which they are vivisected, radiated, poisoned, sewn together, made malignant, boiling water poured into their skulls, etc. The rat on the loose tells them that they should be pleased to be part of such glorious science!

"What you do well," I wrote to Bill, "is create Doctor Rat's voice, follow its obsessions to the tiniest detail, and allow it to grow its own psyche from within. He is almost without a misstep. You find his spot in the cosmos and elaborate it elegantly, chillingly. The true content of Doctor Rat's rant is its almost libidinal anti-libidinal urgency, the cadence that make him (and ourselves) real—the relation between desire and its opposite, meaning and its opposite, madness and its opposite, death and its opposite. Unlike other 'animal rights' documents that were constructed years later ideologically, your parable strikes at the deception that normalizes torture chambers and incarceration of spirits. It is centuries ahead of PETA and post-modern animal-rights propositions in its aesthetic, its primal language, its appreciation of the main thing that animal advocates should be alert to—the antitheses that lie within the dreams of mankind and all species, binding them to each other, that keep returning (beast as man, man as beast) to claim their due."

Bill emailed back: "I visited the labs I wrote about, as well as slaughterhouses. I saw nice young grad students beating a monkey on the head with a hammer. They were very objective, they smiled. Am I dreaming? I asked myself. I came away shaken to the core. I began *Doctor Rat* as the result of a dream, in which a group of animals said they had a story they wanted me to tell. I was living in Canada at the time, in an abandoned settlement. The story had actually begun years ago when I visited the Bronx Zoo and saw a caged eagle pacing back and forth like a Roman Centurion in a black cape. A sadistic woman was flashing sunlight off a mirror into his eyes. He looked at her with magnificent disdain. He was used to the sun in his eyes. Nothing she could do to him affected his pacing. He was scheming to escape. He recruited me on that day and finally, if only in the astral world, I set him free."

In our universe, electrons are in the outer shell. Like all molecules, we come here to change, to be changed. In another universe, elsewhere, electrons occupy the inner shell, and entities come there to know. Things cannot be altered with their electrons buried.

What is this if not a manifestation—if not a signal between domains so far apart that the universe exists solely to unite them?

Overheard from three young girls and a boy with two rubber inner tubes in a lake:
"We're bad pirates."
"No, we're good pirates."
"No, we're bad pirates."
"No, we're pirates!"

The question of what is going on here is too profound for daily reckoning, so it doesn't get asked. Of course, at the most fundamental level, it is the only question asked; our councils, roadways, and machines are the asking. It is asked (and answered) by shopping malls, bus stations, European football, chivalry, mercados, Chris Rock, video games, surfing, Burning Man, and so on. Abstractly, the question has no answer. Asked in terms of the great dance, the dance alone is the answer.

Many have proposed that we are building here a facsimile of an edifice that already exists elsewhere, at the heart of the universe.

At the same time, we are barely removed from the predation of unsentimental fish and well-equipped cats. No wonder so much of our reply is impunity. No wonder answers are inseparable from questions. No wonder primates matriculating from Islamic madrasas are tough, motivated predators. They are at war with the hedonistic nothingness that has supplanted their saints, prayers, and beatitudes. They are defending the temple from the burghers. They are protecting all ninety-nine names of Allah from defilement by post-modern machineries and globalized tabloids. They are patching the goblet from which Mohammed drank.

Buddhism, on the other hand, concedes everything and tells us to be unafraid. Reality and matter are provisional. Thoughts and beliefs are scribblings in water, as water is. Amnesia is destiny.

Buddhist nothingness is not a dread bastion of loss and abandonment; it is a permanent state of enlightenment, of calm abiding—nothingness only in relation to a tapestry of delusions, phantasms made of molecules, solid-seeming, but intrinsically empty, that dissolve at death, along with the consciousness attached to them. What we perceive as "nothing" is the foothills of another kingdom materialized in its own way—the way of unclouded mind, cleansed of projections and appearances—and thus not doomed like those appearances to vitiation. It is the resting state of the universe from which the ghosts and habits of matter commence, from which atomic charges arise sylph-like and become things, from which are launched the suns of thought. In this "other" universe what happens is real—contemplation of isness without tedium or vitiation.

The question remains: why was enlightenment originally disturbed, why a pigeon's feathers ruffled, why molecules of hydrogen bent into molybdenum, why restless thoughts rustling through ontology, creating universes? The answer is, again, the world we know. If we want to understand why karma has impelled locales like this into being, look at the world itself; if we want to know where our journey is leading, look to current mores; these are lives yet to be.

We are figures in our own dreams living the myths we always meant to live, experiencing the doubts that galactic layers contain within them.

Am I sure of this? Of course not.

But that is why I am here.

When interviewed on *60 Minutes* (December 2004) Bob Dylan rejected the notion that he was "the voice of his generation," insisting (as he has for years) that he had no idea where his songs came from, they just appeared, ready-made in his thoughts, "like a ghost was writing them." He proclaimed himself a minstrel, a truant visited by an angel, not a prophet. He doubted he could compose such "love minus zero," "tambourine man" lines today.

On the same program a week earlier Jay Greenberg, a twelve-year-old prodigy who composes piano sonatas in twenty-five minutes and symphonies in a day, said he doesn't revise his pieces because, as he put it, "The music ... comes fully written—playing like an orchestra in my head."

Soon after completing an original piano concerto in G major Mozart was walking down a Vienna street, its lilt fresh in his head, when he heard from inside a filthy pet shop a starling whistling his main theme. He purchased the bird for 34 kreuzer, befriended it, and recorded some of its subsequent songs.

Whatever we are missing here is more than we are getting. The name "Dylan" stands for something crucial, but far different from what we presumed.

Suicide, by the fact that it is possible, provides a clue to who we are. Machines don't terminate themselves. Yet, if you are a living machine, you *can* turn yourself off. You can get out of here. By pure will, you can excise this whole event.

Travel between life and death is not routine. Once you leap from a ledge, discharge a pistol into your brain, poison or drown yourself, detonate yourself in a plaza of civilians, shut off your body by one means or another—how do you get out of *there*? How do the unhappy dead, should they awake in another time, another place, wanting to "kill" themselves again, carry it out?

Many sects believe that the only way out of "there" is back to "here," or some other place. Consciousness is not so much a self as a shape. Life is not the ultimate status of identity nor is embodiment the only form of life. Mind is merely a shifting terminal, a sorting device for vibrations.

When the flow of sensations to the brain ceases, a person may be blotto to himself (or herself) for a few hours or days, but the mind is playing possum, stunned by the trauma of elision from its recent manifestation. Eventually its true basis starts to seep back in and establishes a new conditional reality. The vibrations return on this basis.

Molecules are inherently empty anyway—so much air and charge—they are not the ultimate instrument of being.

It is generally a bad idea to kill yourself, not because it gives away the only thing you have (or perhaps because it doesn't) but because it puts you in a precarious position—i.e., you have gotten out of "here" where things were rough; now try to get out of there! And even if you do.... It is not the way to stop the conveyor belt or tune out the vibrations.

"Here" and "there," "die" and "you" don't mean what they seem to.

No one is ever totally cured of disease. To assume otherwise is a misunderstanding of our biological situation. Each cure of a pathology reconstellates its essential pattern such that a new disease, a new edge emerges to be confronted in the next act of healing. Life is inherently mottled, ever reshaping itself.

To heal is a crystalline attunement; it involves jingling a watery pattern so that it re-gathers itself and evolves.

The Big Bang is a metaphor. It can only be a metaphor. It is not a real event because it has no locale, no context, nothing but the overkill of circumstantial evidence misdirecting the inquest; its hypothetical occurrence outside sentient imagination is, at best, cult science. More accurately, it stands for a state of consciousness in which a "thing" begins, expands indefinitely, contaminates everything with itself. In dreams, which intercalate waking states (and any astrophysics), the Big Bang always ignites again, and (awakening from dream) ends, returning raw consciousness to itself.

The universe could just as well have arisen by spirals and counterspirals of vital energy inventing space and seeping into starry night. Or a greater dimension putting a single finger on a pond no bigger than a Golgi body.

That radiant urn of gases, Saturn, like so many billions of kindred receptacles throughout the cosmos, holds a molecular abundance that could become river systems, trees, ant-hills, schools of fish, birds of prey, mosquitoes, by reference to a different orbit, baptism at critical mass, a temperate sun, somewhere else. The quotient of organic molecules throughout the galaxies reflects the resting potential of sentience itself. There may be too many for a human or other species to count, to conceive of—far too many—but there is no excess relative to the capacity or desire of mind to attach to matter. That is the base equation and always has been. It is the glowing bounty the *Cassini* spacecraft gazed into as it hovered beyond Saturnian moon Dione—the mother planet luminous and golden, a frigid gravitationalized basin of primacy itself.

The aggregate of the astronomical universe and of the things in it is equal exactly to the luminosity of consciousness; it can never be any greater, any less.

A friend with whom I walked to lunch earlier this week told me, over soba/daikon broth and celery-root pancakes, that a British philosopher at Oxford recently proposed there is an outside chance, not a great but a reasonable one—say twenty percent—that this entire universe is a simulation run on information-processing superobjects by members of an advanced civilization, and that perhaps the simulation is failing and the plug is about to be pulled.

"So does this make your chronic depression better or worse?" I teased.

"Probably better, but think about it: my depression may only be software. There is, in fact, no observational way to tell if we are living in a matrix or not. The universe of galaxies and stars, even the Big Bang itself, could be a virtual background, a screensaver for the program in which our consciousness is being synthesized. There might be no sky at all outside the simulation."

I laughed. He was so quixotic and impressionable. One year it was esoteric astrology and a string of abstruse synchronicities; the next it was a dangerous but too-beautiful-to-resist mestizo lover; the following year she was "totally nuts," a stalker; he had a new, choral girl with whom he necked exhibitionistically at parties, and the place to be was the internet noosphere. He had traveled across decades between apocalyptic nihilism and Mithraic moonshine, and back. In fact once, he quit his professorship because a friend told him she was about to inherit a fortune and would support his writing career for the next few years. It turned out she had this news only from a channeler and the likely source of her bounty was legendary Nazi gold. Luckily he was able to get another job.

Dipping his pancake in miso sauce, he parried my smile: "Doesn't it look like a simulation to you? Don't embryos behave like simulated objects, programs designed to create the illusion that we are getting made in stages like Mr. Potato Head inside of one another's bodies by DNA programs? You yourself pointed out how bizarre a plan that is. Maybe genes are subprograms of the main program, and creatures assembled in wombs are someone's sense of humor, the thinking man's 'Grand Theft Auto.'"

"Then what if," I asked him, "*we* were to develop the technological ability to run such a simulation in the future? Would the matrix inside the matrix be any more a simulation than the first one?"

"Good question. I don't know if the outsiders would even allow simulated beings to run their own simulation. You'll have to check the website." He wrote it down on a napkin: http://www.simulation-argument.com. Then he called the waitress to order twig tea.

"Who's behind the simulation?" I asked.

"That's obvious. Our descendants; that is, if we survived long enough to develop an ancestor-simulation. If we didn't, there is no chance at all this is a simulation and only a negligible one that our species will survive."

The moss on stone, so green in this dimension, is brighter after the rain.

I am going on sixty years in this shape, the Moon faithful in its orbit.

The only thing that ever mattered was the mystery and the wonder.

You have a registered letter from a person whose name you do not recognize. Even more of a surprise—it contains his *post mortem* will. As bizarre and unexpected as this is, it is just the beginning. Above his signature he has left you, by name, acreage in the Idaho panhandle: county x, district y, plot number such and such. On this land, you are informed, in a grove of conifers camouflaged by a paradox of stones, is the entrance to underground caverns, a stalagmite temple that is now your property and responsibility. You are entrusted not only with concealing its location—its very existence—but protecting the secret that lies within.

What an unlikely turn of events! You don't remember ever being in Idaho ... or maybe once. Does Idaho even have caves?

As in *Mission: Impossible* you are told to incinerate these instructions upon reading them, for they contain a confidential matter more precious and precarious than a mere grotto— the key to the physics of the universe. One day you will write your own document, pass guardianship on.

Don't immolate the ciphered map prematurely (see Attachment A); wait until you find the site. It is five miles into a stand of forest; as predicted, cleverly disguised by trees, hills, and outcroppings of quartz. Without a codicil of landmarks, you would have bushwhacked right past.

Under flashlight a remarkable subterrane looms up: phosphorescent rooms, stone candles, multiple trellises—but, more to the point, as described in Attachment B, the geomancy of its pillars and portcullises have formed a wormhole to an Earthlike planet in another galaxy. This is the treasure you have been deeded.

Now burn!

The previous guardian was an astronomer and, in notes stuffed in a jug between cited stalagmites, he explains how, by a comparison of night skies, he concluded that the planet in the cave could not be in this galaxy. It would take an advanced astronomer plus some ingenious software to make such a calculation, i.e., to analyze that arrays of constellations could not occur from any vantage in the Milky Way—but let's assume that the intra-cave night is radical in all its major star clusters and proximate nebulae. This particular point is incidental, for a portal to another solar system is sufficiently paradigmatic. Transit between galaxies additionally affirms that creation stands in consanguinity from one end to the other, generating the same sorts of worlds and landscapes from stellar dust, wormholes linking the remotest of them and ensuring migrations between star systems millions of light years apart.

The planet in the cave is as large as Earth, perhaps a bit larger, for its gravity sags limbs and lungs. No way of knowing for sure. It could be a former Earth when the Milky Way was young. That depends on how the nautilus of time is cocked. Evolution is at an archaeozoic level. There are plants, worms, insects, lizards (but no dinosaurs), avians, some very small mammals. These creatures are probably not those things but resemble their stages.

It is nothing short of incredible to slide through a cleft in a dank remote chamber into something so vast and virginal — mild breezes, magenta grasses, sweet rain. At night two very small moons followed by a regnant satellite half the size of Luna ascend by their own metronome. In the morning under indigo skies you explore fields in yolkier sunlight than Earth's. A brook tastes of the salts of creation.

If the governments and corporations of Earth find out about this place, they will dilate the wormhole (if possible), punch their machines and vehicles through, stampede into its world, colonize it, haul off its resources, eventually fight wars over it — delete its future. The knowledge of the land's existence should be limited to, at most, a few people.

The Idaho portal always opens on the same vista, so you are likely to be exploring only one prospect of tens of thousands of exotic flowers and shrubs swarming with insect-sized zooids (some of them quite butterflylike, others resembling small airborne dragons), overlooking a bay of primeval water, beacon for a flock of meta-pelicans and other nesting fowl. You can hike from there in any direction for days, even weeks, before hitting an ocean or great lake, but you do not want to risk getting lost. After all, this place is real; the cries of those red "crows" are final. Even assuming an all-terrain vehicle negotiated through the wormhole (a bad precedent), you can journey only so far, faithfully marking your trail — a gamble over forested hills punctuated by rivers so immense you can barely see their other side.

You could ferry a Cessna through, but you are limited to a tank of fuel (round trip). If you develop engine trouble and ditch, you will have to spend the rest of your life alone on this world.

Aerial surveillance reveals just a cut of the geography. Beneath your banking arc, waves roll to ungaugeable shores. As you scan the primeval waters—occasional flying "fish" and "mammals" leaping in their indigenous minds, a lenticular cloud stabbing the horizon like an emblem of forbidden worlds— you turn back. You have no idea what dwelleth where this sea laps upon its nether brim—mountains, Brobdignagians, Lilliputians, unicorns, thundering surf, fjords, villages?

How can you study this planet without letting the hordes in? How can you explore it without subverting it?

But this is also your choice every day you wake up here. We have reduced our world to global positioning points and strategic landscapes, surface histories of intra-species conflict. We are urbanizing and suburbanizing, paving leas and marshes from Liberia to Brazil. It doesn't take much mercantilism and vigilantism to shrivel a rainforest into a sandy plain, transiting which an extraterrestrial craft might photograph something like what the Mars lander saw where it rolled to a stop within the rhetorical former lake of Gusev Crater. In fact, some presage that, by global warming alone, the Earth will degrade to the status of Mars, that Mars once had fossil-rich seas, Amazons scissoring continents.

It is probably best to keep your own private Idaho, inviting only one or two reliable friends, investigating the immediate hillside and adjacent territories, importing at most a picnic. Just being on a world is a privilege. Travel light. Then choose a worthy successor.

A cell becomes a small animal, and a hive of such animals matriculates into a woman, requisitioning atoms for its figure and synchronizing its homunculi to a cardiac pump. The profundity of this ordinary daily event is entirely missed.

In 1919, aboard ship in the Atlantic, a young George Kennan wrote: "It is all too rich, too full, this summer day. It is more than one can stand. One would like to cry out to the gods to take it away again."*

He didn't say, as did an aikido master hiking into a Hawaiian valley in 1985: "It is so beautiful I want to die."

I take them to mean that it is overwhelming to be in a body. Our obligation to an unseen presence, to the perplexity and lavishness of manifestation, at moments of lucidity, is too much. Freedom in the bosom of profundity is near unbearable.

Along Arizona's Route 66 at 5,600 feet one encounters sudden winds, microbursts, dust and sandstorms, hail and raindrops so fat they sting. When matter is this active and incontrovertible, the rivalry between matter and spirit gets resolved in favor of spirit.

---

*Ronald Steel, "George Keenan at 100," *The New York Review of Books*, April 29, 2004.

Is this horrible? Or is this wonderful?

Guatemalan shaman Martín Prechtel recounts the day that a paramilitary goon squad, three "short-haired, hatless, tall white men ... in polyester shirts and slacks, cigarettes hanging out of their mouths," invaded the Tzutujil Mayan sacred lodge, "fierce-looking machine guns in one hand and clip belts over the shoulders,"* and demanded to see the most ancient and revered medicine bundle, the Heart of Food-Water. They had heard it was something special and wanted to check it out for themselves. Once inside the kiva, they would insist that the priest unwrap and lay its contents before them. They thought maybe it held gems or precious metals that they then could steal at gunpoint.

Martín, an initiate of three years, was appalled when Chiv, the shaman, and the Rain Priest, sharing a wink, invited the defilers into the lodge, banishing the legitimate students. After a blanket was draped over the doorway, mocking laughter arose as the fatuous intruders viewed contents that devoted aspirants, milling uneasily without, would not be allowed even to look at until four years of arduous training had been completed.

Their initiation had been cheapened, ruined for them, as the commandos strode away, chuckling among themselves at the trifle of supposed treasures.

For what was Martín putting in long days of service and labor? A few beads and trinkets?

---

*Martín Prechtel, *Long Life, Honey in the Heart* (Berkeley, California: North Atlantic Books, 2004), pp. 69–74.

The ceremony was lost, its happiness gone. "We'd been violated by what didn't love us, or comprehend the delicateness of what it took to get to the common spiritual ground where what was human could feast with the divine. The beauty had been scared off like deer in front of sport hunters. The immenseness had been diminished and trivialized. How could Chiv have sold us out like that? How could the Rain Priest have let him? What was wrong with these old men?"

Soon after this indecency Martín led a posse to the shaman's hut where, shunning etiquette, he blurted: "How could you let those *Mosi'* see the Bundle? . . . [T]hese men are uninitiated, they know nothing about the true meaning of the Bundle at all. How could you just open it like that without any of the respect and time it takes to do it right? How come you sold us out?" (The suspicion in some circles was that the old mage had been bribed.)

But Chiv was happily preoccupied. He had been cooking patiently, and he offered his intruding disciples a delicacy in progress. Five ducks had been smoked over hardwood, cooked in "achiote, tomatoes, squash, seeds, and salt, with little chinks of piech', wild banana-stalk hearts, and gourd tendrils sprinkled in," then wrapped in a plantain leaf, tied with fiber, and steamed for another hour in a clay pot. As he enticed everyone to this feast (. . . "first it's good to eat. . . . Look, smoked ducks, five of them"), he told them to chill:

"This bundle is older than any of us, and was here on the earth before any humans were created. This bundle's holiness was not created by people physically, and, for the same reason, there are no people who can destroy it by taking it apart.

"Don't you think the bundle must have its own protections? The power in the bundle makes us whole but we don't make that power.... [What] makes these disrespectful people so powerful that they can diminish the power of the bundle?

"... [P]eople from the outside want to take apart our bundles to see what's in them. You can take one apart in a second, but you see nothing. The only way to 'see' what's in the bundle is to learn slowly how to put it together and how to take care of it, like an egg, for instance. If you want to see why the mother bird thinks her eggs are so precious, like our village loves its bundles, then you could break the egg. All you see is mucus and yolk. But if you initiates sit, hatch, maintain, and care for the egg without breaking it, you will see in the end what an egg is all about. An egg is really a bird, and a bird can lay another egg. To hatch an egg and raise that hatchling so that it can fly takes time, care, and worry. That's initiation.

"By looking in the bundles those people saw 'nothing,' and, seeing nothing, they took nothing away with them....

"... by going the four years you learn to see 'nothing' in a substantial way. What is in the bundle is in the seeing. We are not so primitive as to think that the bundle is the power, it is simply a home for the power to be seen. The bundle is a throne, and it takes four years for our poor human eyes to see the spirit sitting there.

"All of our bundles contain ordinary things that, when seen in a ritual context, become the extraordinary things they really are."

Such is the egg our species has been trying to crack open for the last 10,000 years, initially with chert and granite blades, with increasing fervor the last thousand or so—still with crude stone knives, even today, despite 10,000 years of shamanic warnings and Taoist precepts. In the last sixty-plus we have blasted the egg to smithereens and now insolently attempt to remake it by nanoparticles, to tabulate its every secret as we go.

This is the priceless egg that biotechnicians and physicists transgress with blithe weaponry. They see common molecules, fribbles, just stuff. Yet nature is an absolute integrity, in fact a god; it yields only what is; it cannot be threatened or bluffed. In the guise of letting us snatch what we want, it keeps its principle intact—bait and switch, fool's gold, wooden ducks, Humpty-Dumpty, whatever it takes, to lead us astray. Greedy and gullible, we merely seem to get in, to have the run of the house, to make it over as our lab.

The old alchemists knew: without ceremony or eloquence to address the god, however sophisticated or dominant otherwise, we are dull and impoverished.

No spiritual, psychological, or political leader can be trusted as long as his ego is invested, especially his unacknowledged ego, as long as his acts impose his own, even subtly self-serving, self-aggrandizing agenda.

Your existence is what he does not feel, trapped as he is in his own essentialism. The true teaching is not about him getting you through some consensus bramble patch, his version, his arbitrarily strict house rules or else. It is about you learning your own way so that you know where you are. The path is what you integrate into your essence, so you count on it to appear always, even in coma, to make every encounter what it is and what you are, to be with you as your karma is—whatever bardo you are in.

The guru should not become your obstacle by imposing *his* system, his divisiveness and duality, by pretending (to himself) to know exactly, uniquely where you are. He should not, by his blindness to the shadow and the force of enantiodromia,* push you so hard with aggrandized statutes and authoritarian directives that he elicits your psychic resistance to the very spiritual unfoldment you seek (and he ostensibly seeks for you).†

He may intend the best, even be a generous mentor but, if he's not going to find you in his heart and lead you by empathy, by the selfless joy of creation, then (to save face) he is going to herd you irresponsibly into your own, his own problematic, unexamined depths. He will remain clueless about who you are, as he is about who he actually is.

Seeing you as a person, as another sentient being, *really* seeing you, will find you (and him too) and guide you both through this maze, more so than any stringent hierarchy or iconography.

Beware crazy wisdoms and arbitrary authoritarianisms.

---

*The process by which something becomes its opposite.
†My friend Bill Stranger, who is a devotee of an "enlightened spiritual master," warns me after reading this section "not to embarrass yourself by revealing your ignorance of the tradition of devotion to the Adepts in general or your own inexperience of the transmitted grace of a living adept. The guru is not merely a human being just doing the human thing. The guru is the divine radiance itself, made manifest to born beings like ourselves. His or her qualities are utterly benign, but he can't always behave in the sweet, feminine, or avuncular mode (in other words, always be the 'nice guy'), or he will be mistaken and misused as a parental figure. Very great gurus who are rightly revered to this day have literally beaten their disciples or put them in situations where they have been abused by others. In one famous case, thugs found a devotee meditating in a graveyard and stuffed shit in his mouth. A guru has to do whatever is necessary to show you his and your divinity, to reveal to you that it is possible for the human to be divine, to disillusion you of the entire course of your ordinary destiny, with all the implications of that."

Beware the teacher who knows everything and can elevate or heal everyone. The right to lead and instruct others is a privilege earned by selfless deeds. Beware the master who will not permit his disciples to swim in his private lake or, for that matter, walk near it, though he is out of town forty-eight weeks of the year. Affectation, charisma, and reincarnational bullying are not real teachings. Many a priest, while issuing extreme unction, has depicted hell-fires to a dying spirit, tortures for eternity that can be escaped only by rewriting his will at this last possible moment, signing over his earthly possessions to the church. This is not spiritual guidance but the oldest con game in the cosmos—locally it hearkens back to the Middle Stone Age.

The key, as the Dalai Lama put it, is simply kindness.

A paranoid schizophrenic subject to unpredictable hallucinations was told by his psychiatrist that if he saw something that shouldn't be there, it wasn't there. Driving outside Washington, D.C., he plowed into a grand piano that had fallen off a truck.

Systems don't work. There is no global positioning device for reality. Only *you* know what's there.

The same holds for the likely apocryphal Kansas woman who purposely stepped in front of a moving car when she saw inflated dolls ascending into the sky after a traffic accident that toppled a pickup truck of adult toys. She thought it was the Rapture and wanted to go to heaven too. Ditto those biblical enthusiasts who scheme apocalypse in the Holy Land in order to hurry the Second Coming, as if history were God's roll call.

Guess what: God created beings not to act in a morality play but to experience what is unfathomable, to elicit what can become, to descend into the darkness of creation and reveal it to him, to mourn and celebrate enigma and possibility. The universe is a whirling dervish, not a hanging judge in robes.

There is no formula for cosmic instruction. Hyperspiritualized rituals are more bling bling in the global market, a ceaseless purgatory of products and buyers, slogans fluttering through bardo realms. No pantheon of angels, demons, or tangkas, unless fully internalized and made one's own, can serve as tracery to an actual place. All are mirages or forgeries, except for the living.

My stepmother, Bunny, used to say, long after her sons had grown into adults who blamed her for their problems, "I couldn't give them what I didn't have myself."

This is deceptively profound. None of us can give what we don't have. Not therapists or teachers, not politicians, not even lovers.

Polemic Two

The Bush doctrine—spreading freedom and democracy indiscriminately across the planet—has no moral imperative or legitimacy because it is being imposed in the context of (otherwise) a brazen transfer of resources from the poor to the wealthy, plus selective emancipation based on geopolitics and a program of aggressive environmental plunder. One suspects the sincerity of a leader who, while claiming a mandate to liberate the oppressed from despots and evil-doers, simultaneously supports a form of globalization that makes the rich and powerful richer and more immune to oversight while further impoverishing and disenfranchising those who can barely survive as is; who encourages—even flaunts—high-pollution industries manufacturing shamelessly insipid commodities for exploitation of both laborers and consumers; who proudly engineers corporate takeover and stripping of public lands; who facilitates plantation slavery managed by transnational bureaucracies across Africa, Asia, and Latin America (rubber, sugar, lumber, fruit, rice, diamonds, etc.). You can't steal from the poor and sell them back their own staples and native seeds at whatever price the market will bear and, at the same time, pretend to be some sort of super-tough good guy. Either you are a man of honor or a mountebank. George W. Bush's piety fails the basic smell test.

On what spiritual authority do you preach a culture of life while ignoring the mass casualties of the wars you instigate with tawdry eagerness, after presiding over record numbers of government executions with the self-righteousness and reprisal of an ayatollah? On what moral basis do you present yourself onstage to humanity as a crusader for human dignity, a redeemer of torture chambers and liquidation squads, while backstage redefining the Geneva Conventions to allow interrogations by stripping, hooding, sexually mortifying, religiously degrading, coercing with dogs, forcing sodomies, and hanging from a ceiling and beating to a pulp (while laughing at anguished cries to Allah)—acts for all intent and purposes bearing your signature and smirk—routinely carried out at Guantánamo, Abu Ghraib, and the Bagram Collection Point in Afghanistan? If these things happen on your watch under your tacit orders, you cannot just dismiss them as errors by overzealous soldiers of low rank; at very least you must show real regret, do public penance, and deliberately reverse the protocols. Otherwise, you are hypocritical and cynical. If your person and policies betray no compassion, then you are probably not compassionate. If you instinctively prefer lethal injection, no-holds-barred cross-examination, and the stampede of tanks and targeting of missiles to diplomacy, clemency, and the Golden Rule, then talking "church" and "life" will not reclothe you as a man of peace or messiah for God.

From Egypt to Rome to imperialist England and post-communist America, public religiosity has been a disguise for the totem power of the hereditary ruling class, a constituency that cares little for the actual sacred or divine. When overpopulation, carbon combustion, and resource consumption are undermining, at an increasingly accelerated rate, the very basis of nature to support life, when the United States is the singlemost egregious abetter and beneficiary of this holocaust, then the CEO of that polity should at least demonstrate awareness of the problem and take some responsibility for it. If instead he stonewalls and shrugs as if nothing is amiss—no global warming, no depletion of nonrenewable resources, no cavalier cronyism—if he advocates business as usual (and more), encourages profligacy in the service of generating greater capital and sovereignty for the affluent, then he is either clueless or corrupt and unrepentant. To shed crocodile tears over Social Security because you are partisan towards the politics at its root, while ignoring the much graver and more immediate threat to the soil, air, rivers, oceans, genome, and food supply, is to fiddle while Rome burns. The Social Security fund may fail by the middle of the century, but what about the biology of the planet?

Mr. Bush is a class warrior, nothing more. He was raised to defend the aristocracy, to create palatable excuses and smoke-screens for its voracious appetite and self-perpetuation. He can offer no moral leadership because he has no principles, only a patina of faux integrity and trumped-up virtue he learned how to project in smug retorts as his birthright. It works: enough people believe that he is a champion of freedom to put him in office, where he can function as a mafia don, delegating spoils and punishing enemies.

There is a global crisis on virtually every level (economic, social, political, ecological)—and that is bad enough. Almost as terrifying as what is happening to my world is the fact that the citizens of this country have elected someone to lead it who either doesn't understand or doesn't care, or both. Maybe they were duped, but that is not reassuring either. The chief of the most powerful tribe on the planet is advocating trivialities, diversions, and deviant, dangerous things—more consumption, more fundamentalist sanctions, more false bravado and public arrogance, more empty sound bites and lies, commercialization of everything (just because it's there), no support for the inner life or service, prayers solely to idols and commodities. Not only are we living in a civilization in lethal demise, but its leaders are dragging it down ever faster. Our lives bear the weight of not only burgeoning ecocataclysm and economic meltdown but, adding insult to affliction, the government's sleazy deportment, reprehensible demagoguery, and snooty condescension. Every day we have to join in playing "the emperor's new clothes," in pretending here is an ethical nation, a humble democracy. This fabrication darkens everything we do and hope for. We would like to be living in a place where things are well managed and improving, but at least we want candor and right intention.

If the president were as in thrall to alternative technologies, fuel efficiency, and the evolution of post-petroleum cars as he is to the oil moguls, lobbyists, and their service industries (piling obscene billions in tax credits into their already-bloated coffers), then we might stand a chance of ending our total subjugation to that drug before it overwhelms the atmosphere, impoverishes the treasury, and sets the armed powers of the world against each other in an Armageddon to control the final barrels in the last reserves.

If he cared as much about stewardship of the seas and fields as machinery of war—stuffing the Pentagon with the majority of the national budget—then he would have moral leadership to bring nations together against terrorism and weapons of mass destruction, plus he would contribute more truly to a democratic Middle East.

If he extended the same custody, dignity, and protection to the living creatures and impoverished of the Earth that he disingenuously metes to prehuman clusters of cells and brain-dead invalids, if he truly supported old-fashioned families and a level economic playing field as much as he does corporate profits and worker subservience, then he would have moral leadership to inspire the nation's values and encourage faith-based initiatives.

If he troubled as much about leaving a vital, healthy planet to future generations as he does about enforcing superstitious evangelical codes and obstructing gay and lesbian lifestyles, then he would have moral leadership to trim overgrown forests, promote human dignity, and speak for the rights of the oppressed in other sovereign nations.

If he expressed as eloquent concern for the unequal balance

of wealth and power in the world as he does for killing jihadists and suspected jihadists and calling out puny dictators, while enhancing his own nuclear shield, he would get more cooperation on matters of security, enriched uranium, and the like from China, Pakistan, and even Iran and North Korea.

If he demonstrated any concern at all for the national debt that he is piling on the under classes and future generations—the price of his appetite for military adventurism and the corporate trough—while exporting their jobs and selling off the very infrastructure of the capitalism he venally touts, then he might be somewhat believable in his attack on large government, his pep-talk displays of "old boy" common sense.

If he addressed the planet as it is rather than a sham arena for selling his schemes and manipulations, we might have reason to hope and to feel a little better about ourselves.

He might not accomplish anything more during his term, but he would have moral credit to spend. As it is, Mr. Bush is in way over his head and in complete denial, publicly anyhow. He is trying to get his way with a simulacrum of morality, a charade of throwing himself into solving big problems, while bullying people into accepting his puppetry as the real thing.

He brings a "What, me worry?" patina of facile indifference to every crisis. "No child left behind" has the same glib ring of inauthenticity as "your son died for a noble cause." To the question, how can a man who rubber-stamped so many executions in Texas identify himself, through one comatose woman in Florida, with a culture of life, he explains belligerently, "It's the difference between guilt and innocence. They were guilty; she was innocent." To the Arabs and Israelis, enmeshed for generations, he offers his Bush-family tried-and-true snake-oil "road map," as though that's what they've been seeking in the desert for the last fifty years. On embryonic stem-cell lines, he says, "I've made my position very clear. I will not use taxpayers' money to destroy life in order to save life—I'm against that." If that's his position, he might try demonstrating the protocol to his colleagues at Halliburton, the NRA, and Tyson Foods. As for the coming oil shock: "We need to get on a path away from the fossil-fuel economy." Thanks for that one, dude. In acknowledging problems, he has an uncanny ability to trivialize and dismiss them, like: "I'm no dummy. I'm on to that. It's under control."

With George W. you know what "under control" means—he worships at two altars, and they are the same: corporate technology and corporate divinity, aka "one nation under God," indivisible yes but forget the rest. Between brilliant do-no-wrong scientists and interchangeable theologians and politicians on whom God confers the enactment of his plan, they've got every problem licked (either God'll do it or some Yale-educated physics guy working for Exxon Mobil will do it), so all we have to do is remove restraints on their free exercise of power. This is conflating casually elitist atheistic science with provincially theistic Endtime such that they become bed-fellows under one party ideology and one political charisma.*

---

*Enron under Ken Lay and Jeffrey Skillings was a perfect witches' brew of anti-Darwinian evangelical religion and rigged pseudo-Darwinian science. Once George W. and cohorts scented its inevitable cave-in, they developed very short memories about intimate family ties. One night Ken Lay was staying over in the White House, handing down the gospel of free-market energy—the next morning Dick Cheney was spouting it to the world, casuistically chiding California for its profligate behavior ("We don't manufacture kilowatts here"). A few months later W. was putting major distance between himself and Lay—as much and as fast as he could— pretending

W. is somehow simultaneously God's indispensable soldier in the land of the unbelievers and the loyal lackey of powerful techno-corporate interests who would like to run the entire universe through their burgeoning Godless machine.

The world needs an American president, Chinese premier, Pope, or U.N. secretary-general to stand up and say: men and women, boys and girls, we've got a problem, we got to do something right now about the devastation of our world and its imbalance of power and wealth before it is too late. We got to stop the worship of products and fake realities and cults of personality. It *is* probably too late, but the last scant hope to avoid mass catastrophe, or to engage it with suitable gravity and decorum, rests on naming the crisis at once and beginning the arduous process of equitable reapportionment. You can do this only if you have justice and integrity on your side. Since the guy in charge has neither, and can pretend to no real mandate, he overstates and hyperbolizes his so-called patriotic wars, distracting and wasting enormous chunks of our little remaining time and largesse, mortgaging our already-mortgaged future. This is his legacy to all mankind, to all species and their posterity, his unforgivable sin before God.

---

he barely knew the guy and, from an elevated moral perch, promising to slap such corporate criminals in jail.

If you conflate religiosity, predaceous science, and corrupt politics, this is how it will always turn out—a Ponzi scheme of one ilk or another delivered from the millenary pulpit of phony moralistic capitalism, followed by insolvency, and then equally disingenuous remorse.

Destruction by degrees of impersonality

1. You are swept up in a tsunami; seismic plates shift and your home collapses; the Earth is struck by an asteroid.

2. A tiger mauls you; a bird of prey descends on you; a shark clamps onto your leg.

3. An enemy shoots you; a plane releases a bomb onto your street.

4. A psychopath kidnaps, tortures, then murders you. A giant eye uncurls to reveal agency.

5. You hold a gun to your own head. You leave in the night.

The depth and texture of here is the longing, the nostalgia, for another world, a world we can't see.

The world at the root of consciousness? The world from which we come?

Or the bottomless fabric of nature itself?

The dream leads you down the steps of a museum. You know this staircase descends into underpinnings of its own edifice. There is no bottom to the architecture, yet it is architecture all the same (someone, even no one, built the stairs, their crude subterranean banister; carved out these walls, the railing growing more rudimentary as it descends, until it is less craft than caked lava).

Freud and Jung, each in his own way, understood—we contain the path to the beginning; we are the record of how we came.

When there is no reason to be afraid, there is every reason to be afraid.

This is the story of the return to the planet of origin. It begins when the aliens, your lost homeboys, arrive from the other side of the universe to fetch you. They don't necessarily drop their saucer on a field and hire a taxi to come meet you, although they could do that. The stranger who approaches on the street looks just about human. He could be the telepathic projection of a man or a male instar.* He could also be a real man, related to you long ago, at the inception of the universe.

He communicates in a calm, understated way, low-key but convincing, a droll humor, *your* sense. Perhaps he says, "We spotted you out here, one of us. Come see our planet." Generalized telepathy finesses the language barrier—avoiding any actual vocabulary or syntax, transmitting pre-linguistic logic strings cortex to cortex.

How you ended up downgalaxy on this crowded, imperiled orb is a matter of the deepest karma. Suffice it to say that a few people from the home planet get born for a lifetime or two on Earth: Earth, the troubled place; Earth, the violent one, child of Titans and Nagas.

The way we will write the story ... the invitation to go home is delivered at a moment in your life when you'd not mind disappearing. Your wife or girlfriend just left you. You lost your

---

*a transmolecular walk-in (see Jeff Bridges in *Starman*).

job. Your boyfriend told you—that very morning—he was seeing someone else. A good juncture for a revenge fantasy: "They'll be sorry when I'm dead."

You decide to follow the stranger.

People in your immediate circle will conclude that you disappeared as if off the face of the Earth, though it may take days before they actually notice. Relatives, wife, husband, friends, co-workers will query one another, then call the police.

After two weeks, no clue as to your whereabouts and no money withdrawn or credit-card activity, the cops conclude you are not just on a pouter. No one saw you being hailed by an odd-looking pedestrian, then continuing with the stranger on his trajectory. Or, if they did, their testimony was ambiguous, unreliable: the police assume that the homeless dude camped against the building across the street made the whole thing up for a cigarette and Diet Pepsi. Maybe, to misdirect the search, the aliens abandoned your car in a crime-ridden neighborhood.

You will be mourned. Your wife (or partner) truly loved you. You love them too. That doesn't still your desire to breathe free air, to wake where things are going well and getting better.

There are a number of options for boarding their ship. You could be teleported right off the street—as in "Beam me up, Scotty!"—so fast the homeless guy might blink.

To proceed in an ordinary car to an abandoned farm allows a more gradual transition. You don't have to be demolecularized or elevated by anti-gravity. You could abandon the rental forever at the landing site. There, in a remote field, stacked in complexly floating layers, sits a spherical machine, bigger than any stadium, masquerading as a hillock. Since no one comes this way, no one spotted the colossus—a city that can fly.

Can it ever! An advanced race is advanced because it has known for millennia that circles go deeper than lines. All aboard!

Inside corridors of bubbled glasslike stuff you get introduced to the crew: wise and gentle ones. Then—zoom, out of here!

It is always exhilarating to shoot above the Earth—the pop through the atmosphere—especially in a vehicle with soundless engines. "This planet sucks!"

Escaping the local gravity well is child's play compared to where the ship has to go next if it is to get there before everyone aboard grows old and dies. Plus, we don't have that kind of patience. We need to reach the home planet in about a week, but not by combining acceleration with Einsteinian time dilation and the disorientation that will impose, e.g. we don't want you to arrive back at Earth two months hence to find that four centuries have passed and no one you knew or even their great-grandchildren are still alive. You should return to the same civilization, the same husband or wife, the same friends, the same children who declared you missing and conceded you were murdered, your body tossed in some river or a dumpster. If you land back in history after two or three months, or maybe a year or two, in the world you left (like Tom Hanks' Fed Ex employee who swam from a plane wreck to a deserted island and foraged there until he reentered his century on a home-made raft), you will get to tell your story in context . . . and remit a package (see below).

No one's going to believe you in four hundred years—and frankly you don't want to experience the future shock after the oil runs out, governments collapse, and terrorists and irregular armies take over and turn the whole planet into Kigali. Better to deliver a portent now, when it might make a difference.

I opt for wormholes. Transdimensional tunnels are convenient and plausible; they get you to other galaxies without requiring faster-than-light technology; they bounce from the orbital range of Mars to remote solar systems; they explain (sort of) how you got incarnated on the Earth in the first place, how travelers from another galaxy could be ordinary men and women.

Even with a wormhole, it should take eight to ten terrestrial days—a lighter-than-aluminum disk slicing logarithmically through the cosmic warp.

Romance on board is inappropriate. Yet it depends on the audience you are imagining for. Is this a metaphysical drama or an erotic soap opera? You can try out both versions. If soap, you might prefer sex-positive aliens to fellow tourists. [Poor Earth. Even the most innocent attempts to reach the home planet end up in seduction and shame. Let's erase this segue and go back to the chaste rendition. It is a dangerous moment. You can debate only for so long what kind of journey to undertake, who your shipmates are, what transpires en route, before you get bogged down, bored with each of the fantasized possibilities, and drop the account, to start over at another time, if at all, with a different plot—perhaps being picked up by a galleon of returnees from many worlds, or escaping to the home planet for good (though that won't cut it because, in the end, you'll still be here). Since the mind tends to flirt with endless variants, in order to reach the home planet before it evaporates, you are going to have to discipline yourself over this rough patch. Otherwise, the encounter never will have happened.]

You have gotten there at last; everyone is excited about your visit because years ago they detected your spectral glow and set their laser scopes on you. You have been their "Truman Show"!

You want to keep the aliens humanoid upon arrival. To have them drop telepathic screens and become giant birds or bespectacled lizards would be devastating. You don't need to wonder how you could have forgotten a whole quadrupedal lifetime or the ability to levitate. You also don't want to have to worry that their entire story was a ruse to get you to accompany them and you'll spend the rest of your days in a zoo with prisoners from other galaxies, ogled at, even in a fantasy you control. (Talk about the perverse vagrancies of our "compulso" minds!)

Since the home planet runs hundred-percent ecological industries, from whence do they generate so much clean energy? How can factories chug away for hundreds of thousands of years without exhausting local resources or desecrating the environment?

I would give them a high solar (photon-psi) technology—the output of their proximal star converted by various modes of telekinesis into everything they need. No animals need be slaughtered, no plants harvested. All potential molecular waste captured by group telepathy is conducted to where it is most efficacious. Few atoms can be squandered anyway when you must get through interminable industrialized eons without self-toxifying or eating yourself down to the rocks and soil. Bumperstickers on Earth warn, "Dams are temporary; extinction is forever." Time is our puzzle and enemy; time is *their* teacher.

Having solved the conundrum of anti-gravity, the home-planet avatars run post-Newtonian, post-Heisenbergian machines, for they are not limited to our thermodynamics; they apply laws four through twelve ... are investigating beyond. They operate their mills along a cusp on either side of which energy is matter, *vice versa* more so. They understand the relative basis of their own solidity as well as the primal manifestation of heat. This is how they tracked you, on a different world in a different body.

Remote visioning and telekinesis sound difficult but, once you get the knack, molecular projection teaches itself. Early struggles locating something even a few yards hence by your mind and then awkwardly budging it mature gradually into a capacity to feel along and search the entire universe.

There are no real boundaries, atomic or other, between self and not-self; it is all heat and intention. A ripple expands, becomes multi-dimensionally sensate, sprouts amoeboid rivulets, vibrates in feelers across a room, anastomoses through a house and out walls, then radiates with growing confidence beyond the city like a psychic octopus, an octoblob extending fractal arms along each tendril to a tingly pod, travelling, reaching, reaching, always following sensation, always in touch, across the stars and into their worlds. What initially felt like a faint fluctuating trickle or muffled pulse becomes exponentially more penetrating than a radio telescope, more precise than a cyclotron, stickier than electromagnetism and more seminal than an atomic bomb. On Earth we have not begun to tap sentient potential.

You can arrange your itinerary and sightseeing as you wish. This world has no violence, poverty, crime, or jealousy, but the homeboys and homegirls are busy. They can't just develop quantum artforms, riffing on virtual clarinets all day and tapping into music of the spheres. That's angeldom. With no inequities or wars, they still suffer and yearn. They, like we, are souls in a Darwinian jungle. Catastrophes and personal tragedies, earthquakes and hurricanes, disease and premature death, even famine and predator stalking—despite advanced civilizations on many worlds—govern all galaxies. In a benevolent universe, life hurts, life is going somewhere. The paradox of creation lies *inside* consumption and conflict. Bodies have to be used. Otherwise, existence will dissipate like morning mist.

Utopians labor to convert molecules for their ultra-dense populations without violence or agribusiness. They construct meta-cybernetic factories along regional sites. You tour miles-long networks of gossamer, transmetallic pipes, resembling mobiles from Alexander Calder—a field lesson in non-Euclidean geometry.

Eventually you have to return to the Earth.

I would encourage the homeboys to deed you: 1. a CD, DVD, or other memento of the trip, adaptable to contemporary terrestrial format; this device must be unique enough to be distinguished from the merchandise of Industrial Light and Magic; 2. selected artifacts from the home planet (multi-silicon-silicon-zinc-zinc-antimony articles that are elementally undiagnosable by known forensics); 3. a hint or two toward a breakthrough on solar and/or pure gravitational technologies so that *Homo sapiens*, mired right now in pollution and bingeing, can adjust toward the ways of the home planet. These will have to be practical diagrams, step-by-step blueprints that can be back-engineered on assembly lines at our stage of development, hopefully cold-turkeying our petroleum orgy while granting our descendants chassis more propulsive than horse, sail, or dogsled. After all, the medallions and foodstuffs of international commerce must still be delivered, the ancient potlatch maintained, until we outgrow it by other, post-industrial means.

Without a hasty metamorphosis, things here on Earth are going to crumble faster than you can say Jack Sprat because people are in an "acquisitiveness/commodity-worship" trance;

they cannot halt the consumption scourge soon enough—and there are only so many ways to kill the goose, the one who lays golden eggs, before the sky is as barren as the seas. We need a slow descent into a different economy, a soft landing in a transitional culture. Immense, heretofore wasted stores of latent energy must gradually replace fossil fuels. People must be useful again, not just mercenaries, operatives, or slaves.

Script your re-entry so that it addresses the "cold case" of your vanishing. Avoid melodrama. Understate it: "I'm not dead. No one kidnapped me. I didn't run away to make everyone feel bad. I was taken to the home planet—see, I've returned with gifts."

Good try. It might even work. Yet you will probably fool no one, least of all yourself. It is near impossible, especially in the short run, to get the Earth evolving along sustainable parameters, to make sure the collective CEOs don't abuse the dowry of the planet of origin—prevaricating to the public that it is monitoring their safety while locking patents up in corporate monopolies. Look at leadership in every nation now, even the so-called good guys. We have bought ourselves quite a cabal, and they don't intend to let free elections slow them down this late in the game, not when they have multiple ways (from bullshit to the push of a Diebold button) to steal the vote. Maybe when telekinesis is taught in kindergarten, food and energy will be as free as air. Then we won't have to dream about a home planet.

You also need to get back into your life without becoming a prisoner of your own story. Give the adventure up as though it never happened. Anyway, it never did. Go downstairs and get a snack; chill; sit on a zafu; breathe.

Soon enough the sun is shining; birds are calling in languages more ancient than mankind, indigenous to the planet; layers of cumulus are sailing to the horizons. They are laughing at you as they puff out their bellies and race at no apparent speed across the celestial plaza. Ravens on trees cry to one another; they don't care about another home planet; they don't even care about the Age of Man.

Scuttle through your given life as a crab or butterfly, easy come, easy go—a gift from the gods, a gift back to same. Leave super-big problems to the avatars—they are better equipped. Or submit them to collective intelligence. The Akashic Record of this world will release what is needed when it is time, skill by skill, revelation by revelation. It will always be a cliffhanger.

Now there is no home planet.

Welcome home.

Out of the prism of eternal nothingness comes submolecular integration. Parts congeal and synergize, small ones first; later, composites. Fish wriggle out of congeries of cells. Organisms accrete in schools, packs, tribes, transnationalities. A city of molecules is erected from the collective mind of fish eggs, an extension of their prayers, shrines, and pilgrimages.

The cougar is a powerful beast face to face, but it is nothing before the corporation—a bag of fur and bones, deflated by a pellet-gun patented, then improved.

We have no power to prevent or curtail the corporation, as the agglomerating features of our existence continue to annex upward, entity to entity, assembling a farrago in our midst.

How many universes will the baby Brahman pipe into existence from his bottomless lungs before one of them keeps, true to his own similitude, congruent with his dream?

Either that, or we pull our sorry asses together and make a go of it here. Remind the caravan that there is nothing but motion and nothing to do but keep moving. No Rapture, no Endgame.

I'm afraid we are condemned to exist forever, as something.

Spiders and gophers accompany us, also forever. Polities made of cells made of atoms, they attend us at every level at which they are inside and outside us. They or their descendants, across five hundred billion years till the death of the universe itself—and that should be more than enough time for them to get us, and us to get them, out of here too.

Orange forms by apricot or marigold degrees in the encroachment of two absolutes: yellow (sun) and red (blood). Some languages do not have a unique word for this state.

Sun, in whose petals I lie; without your morning stroll, no life would exist. Without your immolating furnace, the curtains would not part, a sprig not bud, the fingers of goddesses not trim the arras to sponsor affairs of men. Without your esoteric intelligence we could not write the great book or find our way home.

The ancient power of the Moon was not only that it appeared, juxtaposed and congruent, in the totem sky above primordial farmers and astronomers as the Sun's relative twin; it was that

Luna bore, mysteriously, a different quality of light—a con-current radiance, silvered from gold, at a frequency already half-transcribed to text, an ashen esoteric luminosity in the otherwise starry black by which rivers also swelled and sub-sided.

Pre-technological men and women could not have known that the Moon was an opaque rock mirroring sunshine, one of trillions, a reflected alien day on a tiny unminted coin; but they intuited the solar wisdom inherent in its parchment.

The Sun was too large to read or understand, is still way too large—a ceaseless radiant blood against which the 700 veils of the Moon are hung to protect us from being shattered by essence.

The Moon is a book that *can* be divined—what it says is: "My body is the psyche of creation filtered through the bedrock of manifestation. I am the message of the dreaming Sun made readable in the obsidian of not-Sun. I am a lamp in the bot-tomless cipher across which diadems of other, unknowable suns and their arcana twinkle. What makes me colossal, in scale makes them millions of runic flints."

☾

Spill. Might as well spill.

In 1968 when Lindy and I were living in Ann Arbor, the poet Gary Snyder invited us, spur of the moment one Sunday, to accompany him to a reading at the Detroit Artists' Workshop, a legendary venue for literature, experimental jazz, radical theater, and the like. Our host that day was John Sinclair, founder of the Workshop and co-editor of *Guerrilla: Free Newspaper of the Streets*. A stocky, bearded White Panther honcho in the Ché mode, he was also a bright, well-spoken guy.

When we made a second trip, this time by ourselves, to read with the gang, John took us under his wing, introducing us to Allen Van Newkirk, a cryptic poet whom he described, to his face, as the intellectual force behind the Workshop and, surprisingly, a fan of my writing. I quickly became a fan of Allen's writing too and, over the next year-plus, we hung out together. A Dadaist with a tousled blue-collar look, he was quite tall, sharp penetrating features, emotionally complex and playful, tight intelligent face, the aura of northern Europe. He could have been the nasty but good-hearted lead in a Goddard film. I noticed his trademark motorcycle on the street but never saw him ride it, as we were usually walking somewhere or sitting on the floor, gossiping myth and Black Mountain lit.

A few months later I heard how he disrupted a Kenneth Koch poetry reading at St. Mark's Church in New York by charging the podium while shooting a pistol filled with blanks (though the audience didn't know that), yelling, "Death to bourgeois poets!" His collaborators, among them Romanian poet Andrei Codrescu, tossed copies of *Guerrilla* into the audience.*

I never saw that side of him. With me he was attentive, respectful, bashful—almost protective—of both my writing and personal cautiousness, always eager to share his latest work, no recruiting or haranguing. Political theater just didn't

---

*Peter Werbe, "Former Cass Corridor Provocateur Wounded Following Toys 'R' Us Heist," metrotimes, www.metrotimes.com/editorial/story.asp?id=7169, January 5, 2005

come up between us. I knew that he agitpropped other poetry readings and media events, e.g. the Underground Press Syndicate conference in Madison (shouting, "All media are lies"). He even got his buddy John Sinclair good and mad by denouncing his MC5 rock band, darling of the Beatles and Allen Ginsberg, as "a commodity with a revolutionary veneer." He had a way of getting to the heart of things, talking the talk long before the words were fashionable.

Lindy and I moved to eastern Maine in 1969 and, less than a year later, Allen unexpectedly and abruptly fled Motown to a site even east of us, Antigonish, Nova Scotia, choosing the tundra over the city, ecology over politics (or maybe *as* politics). There he founded a radical environmental center, Geographic Foundation of the North Atlantic, and began issuing literary magazines and broadsides under the logo *root, branch, and mammal*. He sent me copies, and I mailed him our journal, *Io*.

Though our contact dwindled after his first few years in Canada, I honored his impulse when, in 1974 in Vermont, we named our fledgling publishing company North Atlantic Books in solidarity with his Foundation. I wrote to tell him; he thanked me with his customarily humble formality and accepted my invitation to participate in a one-day bioregional seminar at Goddard College where I was teaching with one of his ecoradical heroes, Murray Bookchin. We spent that idealistic evening, long ago, tossing ideas back and forth beside a fire. He was off in the morning, after which I didn't see or hear from him for twenty years. Then, out of the blue, I got a letter postmarked Bellingham, Washington (March 27, 1995)—he had tracked down my Berkeley address. After that communication (see below), which I answered enthusiastically—nothing.

I read in the on-line press that, upon becoming a Canadian citizen, Allen was arrested in Washington State for felony larceny in 1990 and then disorderly conduct and theft in 1994, both prior to the letter. An assault charge in Penticton, British Columbia, led to his jumping bail and going on the lam. He dropped out of touch with folks from his Detroit days except for a rare late-night phone call or sneak visit across the border to his family.

On December 12, 2004, sixty-four years old, he drove a van up to a Toys 'R' Us outlet in a Vancouver, B.C. suburb, headed to the checkout counter, and, at gunpoint, robbed the cashiers of (reportedly) a few thousand Canuck dollars. Pursued by Royal Mounted Canadians in squad cars, he crashed his van head-on into one of them, jumped out, and fled, firing his weapon at the cops. In the shootout he was winged several times, though a police spokeswoman described his wounds as "non-life-threatening." All in all he rang up two counts of attempted murder, one of robbery, and assorted other charges.

"I was wondering what mischief Van Newkirk was into," Codrescu commented to a press guy.

"The surrealists were his big inspiration," Sinclair mused to a reporter at his home in the Netherlands. "If sticking up a Toys 'R' Us isn't a surreal act, I don't know what is! I wonder what the message was. There has to be some subtext." (Other than, that is, the string of puns and synchronicities: Van-couver, Van Newkirk, a van, and "Motown" Sinclair living in the Van homeland.)

I pulled out my 1995 letter today to see if it held a clue, but I have to say, Allen, if there is a subtext, I don't get it. Why? Why? Wasn't there always another way?

"I can still remember talking to you in your living room in

Vermont. The subject that most interested me was your reference to *Hamlet's Mill.* Curiously, I still have never read *Hamlet's Mill,* though I have spent considerable time on associated themes. Let me explain. You will remember that Jane Ellen Harrison is interesting, though not necessarily for her conclusions about mythology. It was her idea that the images or paintings on some pottery, as I call it, are older than the mythographic sources. As it turned out, I probably discovered another technique, not vases or paintings, which has permitted me to walk around some ancient sites and speculate about their knowledge, usually in a scientific art direction. I still plan to look at *H. Mill* and see what is there someday.

"... from the point of view of scientific art the discussion should move from ecology to biogeography. I implied that any discussion of ecosystems would likely result in social science, rather than scientific art, and events seem to have confirmed this observation.

"There are two bioregional series of ideas: Bioregional 1 and Bioregional 2. Bioregional 1 is a formative scientific art theory which takes its origin from my poem entitled '10,500±[Years] into the Nearctic and Palearctic Holocene, with other references,' and Bioregional 2, which is the 'little state in california' notion that is attributed to Berg and Dasmann in the essay published in the *Ecologist* without sources. If we delete the referent you will see that there is little or no relation between these concepts.

"I have outlined an autobiography about my encounters with science and art. It is not an autobiography of people or events, and no references to the 1960s appear there unless they lead in that direction. It unfolds on another landscape.

"[There were some affidavits I prepared years ago. One of

these was entitled 'Affidavit on some discarded fragments from 1968,' concerning the attempt of Orpheus to rescue his wife Eurydice, outwitted by the god of the underworld, as he turns to look at her.]

"The first idea I had in scientific art I called paleocybernetic research, which is more formidable than you might think if you take into consideration the origin of the term. The difficulty I experienced with my thinking at the time was that I didn't know how to form my ideas into a general perspective, and also I was unable to find anyone of the many I knew who would agree that art and science could be combined in this way. I long ago overcame this limit."

For all our knowledge about molecules, birthplaces of nebulae, three pairs of double stars encircling one another to illumine Castor in Gemini; for all the dark matter, black holes, and gas giants, we have not moved one iota from Plato's intimation that when we look at the night sky we are gazing into another world. Not just the immense hydrogen universe of which we are part—not even the radioactive sphere of cosmonauts—another world entirely! This is why night touches our hearts.

The instruments of astronomy have not progressed beyond the astrolabe. The real mystery will never be engaged, let alone solved, no matter how much energy and matter we dissect, how near we project to the hypothetical Big Bang. Physics must address who we are, where we come from, and where we are going.

There is no "history of everything," no explanation, in fact, for anything. It all just is. Cosmology is not about the death of the Sun or the ultimate contraction of space. It is not about a cavernous bowl of twinkling buoys. These are a remote refraction of the true luminosity.

The glow/throb of creation is rooted in being awake and real and, at the same time, totally unconscious, azalea and honeysuckle perfuming the air.

Then comes night, with its zodiac of wandering worlds.

The constellation-riddled hide of the abyss is a giant illegible clock. Astrologers attempt to decode it, to discover its ultimate hours by using what is visible to get at what is not. The houses and signs of various zodiacs computed in towns across the cosmos are inklings of how real chronology is kept among the greater galactic wheels and gears. They all have circles and horizons and linear algebra because a rotating object is wound along its axis into star fields.

Astrology may be superstitious quackery of the most sublime sort, but it is an attempt to tell the deep time of the universe, to read cycles and synchronies that are far subtler than measurements or calculations of mere physical stuff, to track Troilus and Cressida and other fortune-seekers beyond their human mien.

The macrocosm is a metaphor, a hitchhiker's guide, not an algorithm or a machine. This is what the cockamamie pseudoscience reveals. Its truth is its fabrication; its chicanery, its verity. Only a star chart reveals the nonlinear, acausal basis of human destiny.

A Bad Line to Use on the Universe
"What are you going to do, shoot me?"
Especially when the universe is a nineteen-year-old mugger
with a gun.

A complexity theorist collaborating with a rocket buff invents
a totally new means of propulsion out of which they fashion a
spacecraft that, sweeping up particles of dust and fissioning
them as it goes, can approach the speed of light without harm-
ing its occupants. Its acceleration changes its material nature,
which in turn increases its acceleration.

In the new media technocracy the inventors got a chunk of
their funding from HBO, so the spaceship debuts on a six-
show special featuring classic mysteries of the outer planets
and their moons.

The trip to Mars takes twelve days. With an international
audience as their witnesses, the crew overflies one of the regions
distinguished by transparent worm-like tunnels. After they spot
a prominent series of glassy tubes, the producer tells them to
descend. They land cautiously on the edge of a gigantic ravine.
On site, the constructions betray that they are manufactured—
polished rebarred sewers—but where are the ruins of their
cities? Is the Red Planet merely the largest extant chunk of an
urbanized world that exploded three million or so years ago
to form the asteroid belt? Or is it that planet's orphaned moon,
blasted on one side by accelerated fragments of the detonation
and inundated by its tsunami? Are Deimos and Phobos moon-
lets of a moon?

A pass above the notorious Face at Cydonia shows it is

eroded beyond even simian resemblance, though its collapsed Sphinx rests upon an indisputably artificial pedestal with hard margins. Is it the marker of a doomed race, their declaration to us that we come from the same ancestry?

Landing nearby, the explorers soon discover, around the object's polygonal base, alphabetic characters that look nothing like any terrestrial hieroglyphs. Within a few days those cuneiforms from long before the Age of Man and Woman appear on freight trains, posters, and T-shirts across the Earth. They become pop talismans.

What a *tour de force!* Without some quantum engineering breakthrough shortly, leading to an equivalent expedition, neither the glass tubes nor the Face will be explored in the twenty-first century. In fact, under accruing physical, social, and economic constraints, there is no guarantee humans will ever walk on Mars.

Since interplanetary travel is daunting, other scenarios come to mind: Sometime in the next several decades a Martian rover scoops up a divot encrusted around dozens of sleek objects that look like tiny CDs; astounded scientists go into overdrive. Are these the relics of a Martian library? Could the rover have had the uncanny luck to hoe into the site of a Martian Blockbuster video store?

Clearly NASA (or some space agency) will send a follow-up lander, but should it house sophisticated equipment for reading the CDs *in situ* or be a retrieval craft for collecting the archaeological dirt and conveying it and its treasures back to Earth?

If the gizmos are what they look like, the options for getting information off them are greater with the files in hand. One would presumably be able to back them up and then proceed step by step along simultaneous protocols, employing a range of strategies and formats well beyond what could be flown as a payload to Mars and executed at a distance. And this does not even take into account the hazards of a robotic limb getting the disks out of the soil into a computer by commands issued from millions of miles away. How to sift the dirt and scrub the artifacts? What if the device ejects them as unreadable under all superventions? Let's go for a recovery mission instead.

CDs—plural? All it would take really is *one!* Maybe a few dormant seeds from a long-extinct Martian plant would cause almost as much of a paradigm shift—and carry even more data.

With manned flights far in the future, salvage has to be robotic. It will take incredible precision to land a second rover close enough to the first to crawl bumpily along the terrain and join it in machine parley, then effect a transfer of goods suspended forever in the arm of the first. Once the CDs are stashed in the recovery pod, it must roll safely to its lander and reattach. Then, by remote command, the probe must ignite from the Martian surface, tear itself out of the local gravity well, and propel unerringly back to the site to which Earth has orbited, all without a gaffe.

In fact, the mission managers should bundle those disks in the closest thing they have to an indestructible case and collar it with an imperishable beacon because, if the ship crashes on take-off or loses its way back to Earth, the merchandise must remain unharmed, trackable by future generations. No one wants to risk losing such irreplaceable cargo—even if there are (perhaps) hundreds of CDs in the adjacent turf. After all, each file is unique ... and yet all it takes is "one" to bring old Mars back to life.

If a rover strikes such artifacts on Mars during this century, several more years will be consumed fabricating a spacecraft with a specialized landing vehicle, an additional two or three to execute the mission which, only if successful, will place a packet of CDs in Earth orbit for direct re-entry or recovery by a safer, more maneuverable vehicle. Imagine the suspense attending terrestrial approach of the rover, the attempted deciphering of the CDs! Will the discs comprise only nonsense series akin to junk DNA? Will they be blank, at least so far as we can tell?

The mere act of unfreezing their data streams will link minds across millions of miles of space and epochs of time. To read their bites will be to speak a language older than planets and species.

It is one thing to watch a recent movie not knowing the upcoming nuances of plot as its labyrinth unfolds but, if these CDs turn out to store Martian performances, we will not recognize the creatures, the landscapes, the foliage, the artifacts, and certainly not the languages and subtexts. For a Martian 3.2 million years ago, what constitutes intrigue? Inevitably, something will make sense: movements that telegraph gestures (friendly, hostile, goofy?), domiciles and their portals, meals, vehicles zinging along glass tubes, love or what looks like love?

Will we recognize war? Eros can take myriad forms, but destruction and slaughter are unmistakable.

Will its "actors" be humanoid? Will we feel catharsis or only weirdness and dissociation?

Can we tell the difference between alien make-believe and alien documentary—costumes and sets as opposed to natives on boulevards? As our own animators morph humans and animals into digital creatures, landscapes into proxies, that distinction is becoming murkier every day.

How long do Martian cinemas run? Do they have sound—or is that just static? Is their orchestration musical to our ears? As ingenious as Bach? With what instruments? As melodically convincing as Andrew Lloyd Weber?

It certainly can't be reassuring that not only are the "actors" long dead and the scenery vanished, that not only are the houses and the cities of the film-makers gone, but there is not a trace of where they could have existed.

Maybe the movies were not shot on Mars but brought there by tourists. But from whence did these travelers come? Why Mars? Are we their descendants?

Turn the tables. If evolved "primates" on a temperate Venus 750 million years from now should send a rover to their neighboring extra-solar world and happen upon the ruins of an Earth video library, would the Cythereans be able to discriminate our courtships from our battles? Would they understand Bugs Bunny as a fictive zooid, a "cartoon" rather than an object of primitive worship? Would they grasp the theatrical and game nuances of MTV and football, or would they take these literally or in some other context?

If a Martian lander uncovers seeds instead of CDs, would we bring these back to Earth and try to grow their plants? If grains, would we introduce their "quinoas" into our agriculture?

Back to the "real" journey. It takes several additional weeks for the crew to reach the Jovian system, as they must be careful not to accelerate too fast through the asteroid belt where the downside of a plethora of fuel is the heightened possibility of collision with a rock too big to negotiate.

Landing on Europa, they pan across impenetrable ice-pack shrouded in blowing snow that near obliterates the image. The temperature is so far below freezing that only a catastrophic explosion of the Sun could bring it up to "just cold"—not a safe terrain for snowshoeing whelps. Instead the third-planeters work their way, slowly hovering over computer-generated grids of the surface until they find what they are looking for: a gigantic steaming geyser, half a mile across and unleashing torrents of water thousands of feet into the black. It is dangerous to approach because of the immense molten splatter and its erratic, cumbersome splotches.

In the faintly lit dawn, using an all-terrain vehicle modified for Europan excursion, two members of the crew approach the outskirts of the splashzone. Scattered across it they find hundreds of carcasses, some still quivering, of poor fish-like and seal-like denizens who have been ejected from their habitat under the ice to this cruel death beneath Jupiter's behemoth gaze.

Next stop, Titan—what a surprise if, beneath its clouds, there should be cities the size of Djakarta or dinosaur-like beasts yawing in ethane rivers 290 degrees below (Fahrenheit irrelevant). How would they have gotten there without enough heat even to spawn nitrogen-converting bacteria? Can there be a cold-life biology parallel to our tepid one?

After the Saturnian system comes battered Miranda, renegade Nereid, enigmatic Pluto.

Everything is exactly what it is, this too: this self, this current of language-making thought, this way of life, this way of death —all what it is, like this. And no one can make it any different, fix or sate it: hunger, attraction, wealth, power, shame, war, desire, danger, flight. Each of these is innate, inextricable, absolute—the way anything gets born, the inevitability with which anything and everything becomes, the passage of its demise and complete annihilation. Nothing cuts any slack. There may be oceans of compassion somewhere in the universe, but not in the assignment of bodies or conditions of circumstance. Whatever we imagine as our lot, we are cast into armadas as surely as ants, set on tracks like squirrels and quail.

If there is enlightenment, it is not a method, a technique, or even a plan. It is not an aspiration or incentive, neither earnest nor carefree, not a place to go or a state to be. It too just is and occurs because the rest of it is. We become enlightened or not, loved or not, fed or not, anything or not, here or not, much as fog . . . as fog becoming rain.

Tibetan Buddhism has developed a science of our condition—not a science of matter, nor a bioscience, nor even an existential psychology. Generations of lamas have uncovered an actual alchemy of being, equivalent to but quite unlike the West's alchemy of metals. It will equip no motorcars or computers because these are ephemeral protuberances, though masters jet from country to country, teaching the deceptively simple practice of "all just what it is."

Educated and conceptual, able to launch telescopes and deploy armies, we are running from what is. We do not grasp or even see the world we think we have conquered. As we colonize space and supervise matter, we retain an inner sense of smallness and restriction.

We do not know how to make space in ourselves for our actual experiences. Afraid of who we are, trapped claustrophobically in little mechanical deeds and inflated tautologies, we have no expanse for the universe, having projected it into an illusion of things and names; into labor-intensive machineries and schemes; into melodramas and entitlements that arise like clockwork.

The noble truth comes without eureka or satori, without fireworks or reward; it is just "oh"—"oh, cool," but nothing changes.

The science of being teaches how to make space in us for what we are. Only then will wars and privations end, as a candleflame will vanish only when the wick and air that sustain it are no longer there.

You must rely on the mercy of the angel. She is infinite and everywhere and extends the justice of God's love for creation. The universe answers *all* prayers, but not literally or explicitly, not mechanically, or it would be a stagnant universe. It takes them into its critical balance, its unfolding design, to adjust moment by moment across its mega-turbulent breadth for the cries and provisions of its tiny denizens, attending factors so infinitesimal and distinct as to be beyond any inventory, any ken. And it takes a long time for the answer to travel from the center of creation to galactic central to Earth, except that (paradoxically) the answer is also instantaneous and beyond understanding.

On Slavoj Žižek's Welcome to the Desert of the Real

The W. team wants us to think that the jihadists have no political position, at least no rational one. They would like us to believe that we are facing gangs of insatiable madmen jealous of our showcase democracy, eager to deprive us of our freedoms, agents of unmitigated evil who plan to eradicate us recreationally.

Not true. Foreign fighters in Iraq and the international Islamo-terrorist movement have a rational agenda all right; they stand for more than rabid hatred of the West and its "freedoms," envy of suburbia, or subjugation to Allah, superhero.

They want America and its libertine culture out of Islamic countries. They want an end to America's support of the imperial Saudi regime. They want the U.S. and its allies to cease mega-arming and funding the Israelis in their war with Palestine. They want the sieges of Najaf, Falluja, Sadr City, and Gaza lifted, as well as subtler sieges of Damascus, Riyadh, Cairo, and Qom. They want justice for Muslims in Kashmir, Chechnya, Bosnia, the West Bank, Indonesia.

America's peace, safety, and prosperity have been bought by disasters and exploitation elsewhere and, though the jihadists do not explicitly speak for the oppressed of the Earth (clearly not, given their Mohammedan bias), they represent the Islamic oppressed. Their strategy is to bring the same level of catastrophe to America that America has delivered for decades to Islamic communities—if not identical events, their equivalents. Osama bin Laden's motto is: Americans should not sleep securely in their beds until Muslims can.

This is not an altruistic posture on the jihadists' part, but it is also not an unaltruistic one, as most of them have given up personal comforts, furniture, weekends, and domestic life—the usual staples of happiness—in order to serve a selfless cause; they will surrender their lives too (and not just, as legend has it, because they expect to find seventy-two virgins awaiting them). Al-Qaeda and Ansar al-Islam may not be even the social-service organizations that Hamas and Hezbolah are, but they are populist responses to American hegemony. In fact, without their anti-globalist orientation they would have no rationale, no recruits, no *qaeda*.

9/11 and the other jihadist attacks represent not a vestige of medieval Islamo-fascism, unable to grapple with modernity, as much as another, twisted face of modernity—the blowback of a "future shock" Islam onto imperialized capitalism and its Trojan horses of "free trade" and commodity-worshipping techno-culture.

Though we almost certainly cannot (and will not) negotiate with jihad (and its decentralization makes effective negotiation impossible to enforce anyway), we can do a number of things that would be significant, even crucial, in determining the outcome of this conflict, the type of world our grandchildren will inherit. Here are our options:

1. We can begin the long, slow process of retooling our economy and culture so that we are not immured into exporting unjust policies and then enforcing them militarily. We can reexamine actions of ours that exacerbate America's conflict with the Islamic and developing world. Even if we choose not to change *any* of them, we can exercise restraint in our rhetoric (for instance, demonstrating that we at least understand opposing viewpoints and consider them credible and moral). We can avoid needlessly and gratuitously aggravating the crisis by boasting, demonizing, oversimplifying, and otherwise misrepresenting the "enemy." We can drop the word "evil" from our vocabulary because it has a way of turning the tables on those who self-righteously invoke it against their enemies.

If we cannot ameliorate our relations with hardcore jihadists who clearly hate us, we can at least try not to preen and pontificate, to incite support and sympathy for them, to sow desperation so that people in the developing world sacrifice their lives as the only way to regain pride and self-respect. The behavior of the United States under the second Bush regime encourages millions of poor and/or Islamic peoples on the sidelines, who otherwise eschew violence, to cheer clandestinely for our comeuppance.

If you were a child in the streets of Gaza or Karachi, whose picture card would you collect: Osama bin Laden or George W. Bush? Who looks like "the man"?

The "liberation" of Iraq followed by Abu Ghraib and the massacre at Falluja was a textbook al-Qaeda recruitment campaign—hours of free nightly advertising on al-Jazeera for Osama bin Laden and Abu Musab al-Zarqawi, playing alongside the Sharon regime's demolitions of the houses of the families, friends, and rumored friends of suicide bombers. Why would Osama want regime change in the U.S. when he can manipulate the Bush Administration into igniting his revolution? His cross-hairs are on Earth 2052, not the 2004 North American election.

Bush has taken a good twenty or thirty years off America's period of grace before it has to defend itself on its home turf. That is how he will be remembered—as neo-Crusader chump if the Muslims get to write the account, and as the worst president in American history, hands down, if we luck out survival after him, four more years or not.

Look at how dangerously debt-ridden we have already become in financing his imperial adventures. Future generations will look back on this binge in outraged disbelief. What were Bush and his cohorts thinking? Did they not see that the bill, incurred in a time of relative (if deceptive) plenty, would come due in an era of impoverishment and famine? Did they not understand that the ultimate price of the Iraq War was the mortgage of their own grandchildren's future to rising superpower China and the oligarchs of the Sahara and Central Asia?

From a practical standpoint, we cannot bludgeon and exterminate all the Islamic fundamentalists and other opponents of the American way of life, especially as (over the coming decades) their exponentially exploding populations, mired in poverty and crushed by global capitalism, reach critical mass. A transformation in our world-view even without a formal cessation of hostilities or surrender to jihadist demands might gradually alleviate what will otherwise become—has already become—a battle to the death between the guerrilla deeds of those who wish to exterminate our parasitic civilization and our own apocalyptic flails to keep the oil pumping, build fortresses around our malls, and preempt their deeds with mercenary armies and missile shields.

2. Five percent of the world's population streaming along highways crammed with SUVs, an empire guzzling a quarter of the world's oil production and consuming an equivalent share of the planet's resources cannot begin to address the real basis of the global conflict of which al-Qaeda and Ansar al-Islam are early fever blisters—in fact, has no ethical high ground at all. In addition, the Bush Administration has made matters worse by underwriting and even subsidizing waste, ostentatious gluttony, and pollution while boycotting international treaties to halt eco-degradation and economic exploitation.

We have the illusion that we have earned this cornucopia because we are godly and well-managed; we see ourselves as shining knights, the hope of humanity, not craven imperialists and wanton despoilers. We condescendingly inform the world that this is wealth we have created, not appropriated. We claim to be the most generous people on Earth. No wonder our hubris rankles tribes among whom generosity is measured not by handouts but the heart itself.

3. We can develop new policies that, while not significantly impacting the popularity of al-Qaeda, Hamas, Hezbolah, etc., begin to present creative, spiritually acceptable, even Koran-compatible alternatives to jihad. Feeding the malnourished, taking the traumatized and homeless into communities, treating the victims of plagues, halting genocides and relocations, placing orphans in families, decommissioning child armies, airlifting relief after natural disasters, tabooing slavery and forced prostitution, dissolving corporate strangleholds on indigenous populations all could provide outlets for an energy that wishes to serve Allah (or God) and seeks missions of redemption and purification rather than accumulation of goods and wealth. America's aggressive consumptionism and vulgar huckstering, its self-righteous attempt to spread ideology (corporate hegemonies dissimulating as "freedom" and "democracy"), clearly cannot be concealed from the rest of a street-smart planet. You cannot hide an elephant at a bazaar. What al-Qaeda has begun, other terrorist movements will carry to our shores, even if every current jihadist haven is bombed back to the Stone Age.

The Stone Age is where they are coming from. That's home.

We are stuck in a double-bind/double blackmail defined by Slovenian philosopher Slavoj Žižek: "If we simply, only and unconditionally condemn [9/11], we ... appear to endorse the blatantly ideological position of American innocence under attack by Third World Evil; if we draw attention to the deeper sociopolitical causes of Arab extremism, we ... appear to blame the victim which ultimately got what it deserved." He goes on to propose that we adopt both positions simultaneously; "each one is one-sided and false." This is not to suggest a shared guilt that cancels out each violent act by its antipode. We should fight terrorism in all its forms, according to a definition that includes American and Israeli terrorism and the "terrorism" of transnational corporations that impose more subtle and insidious "9/11s" on local and indigenous populations. "... [T]he choice between Bush and Bin Laden is not our choice; they are both 'Them' against Us.

"[America must] finally risk stepping through the fantasmatic screen that separates it from the Outside World, accepting its arrival in the Real World, making the long-overdue move from 'A thing like this shouldn't happen *here!*' to 'A thing like this shouldn't happen *anywhere!*' This is the true lesson of the attacks: the only way to ensure that it will not happen here again is to prevent it happening anywhere else."*

Okay, let's see where we get by the tenth, reverential commemoration—9/11/11; let's see what happens before the Guatemalan calendar runs out in 2012.

There is a rumor that someone in Italy received a message before it was sent.

---

*Slavoj Žižek, *Welcome to the Desert of the Real* (New York: Verso, 2002), p. 49.

In this natively magical world the components of one thing are turning into another, ceaselessly, impartially, globally, inherently, but not without going through shit and decay. A mouse does, in fact, arise from rotting straw, but a mouse can also be cloned from the minute snippet of the carcass of a mouse. Rotting garbage and decomposing vegetation breed goldenrod, phlox, and skunk cabbage, bright golden currency out of a mint. Plants become doctors, scientists, medicines. Dump sites are Choctaw nurseries, schools and temples.

From a pile of dung and slag, earthworms and beetles matriculate, larvae of flying things, mites, and spirals of cells morphing into tubers. The Earth is a foggy procreative flask, with solvents, subtle gases, and minerals roiling amongst one another. Everything is utilized or evacuated for other use; everything is centrifuged and dispersed, refashioned and remade endlessly, carried up- and downstream, though the scale of planetwide vitality and its tipping point are unfathomable.

Amidst this flux, there is no trident of spontaneous generation except in the factories of humankind running polished gloss, parasols, neon, whorled glass, and rainbow-infused plastics and aluminums off assembly lines. These charming counterfeits must fade and decay malignantly before they return to the cycle and become again what they are. Rotting blue toys, yellow and red carapaces, purple and teal fibers litter the planet before the winds and tides sweep them up. No fake magic, sham creation is allowed for long. The only real factory is excrement, mulch, and blood; seed, fetus, and placenta; even to make a blade of grass, to burst into orange mushrooms, or arise from an egg with damp, fragile wings, like the universe itself from the great molecule.

At the corner of two major thoroughfares, signals on mineralized stalks direct traffic—first, one boulevard green (the other
red); then a green arrow for those turning from the first into
the second; then red everywhere except for a green pedestrian
WALK; then the second boulevard green (the first red); then a
diagonally facing green arrow.

Combative drivers and walkers are more or less compliant
because nothing is more basic about the world than mass,
direction, distance, and momentum. If you are going to have
something as sophisticated as information-conducted internal-
combustion engines propelling slabs of metals along curving
trajectories, you are going to have to honor "here" and "there,"
"left" and "right," "go" and "stop," and rates of approach.
Otherwise, objects collide.

Words have the problem that, even when they are accurate and
represent meticulously precisioned thoughts, fully vetted in one's
own mind for sophistry and self-dealing, they are only mark-
ers—and volatile, devious ones at that. At core they are animal
grunts fashioned expediently and prehistorically into signifiers
and solemn denotations. A sociopath convinces himself as well
as others that he is a good guy, a righteous citizen; he experiences the facile expressions of a conscience he does not have (we
are all pathological liars; that is the premise of legal systems).

Streams of metabolism and synapses generating semantic
strings do not even embody the sounds they spew. Despite the
alphabets and semaphores into which they were initiated at

the dawn of history and which have expanded into full-fledged fields of meaning, they remain barks, whinnyings, and chirps of inhuman zooids as well as mumble of the untamed universe. The brain does a masterful job of turning such babble into supple logic and making phonemes feel like true things. Sibilants and fricatives, clicks, nasal resonants, glottal and aspirated stops churn an autopilot chatter atop an existential universe that is reinventing itself from moment to moment, while the artificial designations and assignments of cultures struggle to hold onto their signs.

Even the friendly, familiar words that you are silently sounding, recoiling in neuroelectric bursts throughout your brain zone, that will continue in ceaseless self-dialogue after you set aside this text, do not have intrinsic sense. They are symbolized barks descending along a gradient into meaninglessness, alien cries of unknown tongues, nonsense syllables tumbling toward dementia.

Language is overvalued in relation to action. Wrong action is disingenuously justified as long as the right thing is said, to oneself. However, right assertion that is not grounded in right feeling and that does not lead to empathy is blarney or cunning.

Do not wrestle futilely with your thoughts. Change your conduct because, once behavior itself is different, thought will be transformed at its base. Lacking the right words, do real things—new words will be born.

Isaac Bashevis Singer wrote a short story on this theme; I read it maybe twenty-five years ago. Rather than tracking it down for an accurate rendition, I will tell you the version into which it has morphed in my mind:

A guy mistreated everyone, his wife, his children, his relations, fellow villagers. He was ornery, greedy, ill-tempered. No matter how often he resolved to change, he eventually reverted to his old habits and became increasingly more intransigent. His wife brought him to many rabbis (this was shtetl Eastern Europe). They each provided strategies and practices, and he willingly tried them all, with success for a while, only to revert to his nasty, grumpy self. Finally, in desperation, the couple took a long wagon journey to the rumored wisest rabbi in the district, many villages away. Although reclusive, the old man granted the strangers an audience and, after hearing the history, said simply, "Instead of trying to change your mind, change your actions. At heart you are a good man. The actions will teach the feelings. The feelings, over time, will change the thoughts. There will come a day when, without fanfare, you will feel and think things that match your actions."

After their return, people were startled by the change in the curmudgeon; everyone agreed that a miracle had been enacted. No one could guess how, from a brief encounter, the shaman rabbi had effected such a personality transformation. His simple lesson had turned a small-minded grouch into one of the most beloved and compassionate people in his village.

It doesn't matter if thoughts are greedy, misanthropic, obscene, cruel, remorseless, whatever, because that is the promiscuous, savage nature of mindedness. Gossip continually arises: ungenerous judgments, idle curses, assassinations of enemies, irritable exterminations of this and that, insipid barbarities and sterile libidinous fantasies. Do not get mesmerized by these. Do not judge or condemn yourself in advance. As long as we don't know who we are, we will identify neurotically with relentless, stray, preconscious thought-making that has nothing to do with us, and lose our hearts.

Since the mind can think and imagine absolutely anything—and then berate itself for its ungenerous nature (and words are loose-lipped and wild)—it is possible to spend a lifetime fleeing ghosts of nonexistent selves arising from scraps of media headlines and the marketed fantasies of others. Thoughts are a dime a dozen. What is done is real.

In an unexpected attack on the City,* a catapult (well-hidden somewhere) launches immense rubbery balloons laden with pigments. Appearing in sky-scraping arcs, they seem to drop from nowhere. The paint they bear is so syrupy that single fat elastic drops of precipitation change shape and undulate blimp-like as they float from the sky. The bubbles are gentle; they do no harm as they touch, pop, drench, and saturate wherever they land—lime green, canary yellow, cerulean blue, poppy red, mallow pink, milk white, cadmium orange, mauve, shamrock, azure, umber, heliotrope. People, cars, sidewalks, roads, buses, buildings, trees, power lines are bathed in luminescent color.

Our acts of spirit, our zany courage, pratfalls and improvisations, in the face of multiple layers of contradiction, paradoxes within paradoxes, natural obstructions and sheer perversity, tragedy and mortality are what the deep, all-seeing eye of the universe does witness and register in itself, whatever and wher-

---

*Use for insomnia.

ever that eye is, camouflaged in however many gullies and folds, however latent and tangled the labyrinth and blasted the night. Our spontaneous bliss, our playfulness, our compassion peppered with humor and grace, are what the universe acknowledges and honors, takes into itself forever. The rest is slag.

The new porno obsession of men is staring directly at female genitals. Breasts are passé because they are not "down" or objectified enough—it has to be the barrel of the gun. The goal is clearly not eros, or even allure. It is ritual dissociation and enslavement, the siphoning of desire out of love.

In an age of ascending female power and male self-doubt— and the full force of that power is only beginning to enter our world—the true feminine has become unendurable to some men. They want to bow in worship, to submit, to writhe in rage-filled admiration. They want to adore. But all they can do is twist in the breeze and pretend that they are diggin' and gettin' it, that this is way bad and cool and they are *on it.*

Imitations of pimp and gang culture clone each other through a wasteland of recursions, but these inflated travesties of patriarchy-in-decline cannot stop the goddess from dazzling. "Kill the bitch" rap is the fake triumph of the horny marginalized male, who has lost his pretense to heroism along with any samurai calling, as he is subjected to unrelenting seduction by something that feels like sex but is gamier, more ravishing, and more mysterious.*

---

*As a masculine transvestite on *Taxicab Confessions* put it iconically, "Men are such pigs."

So-called g-string divas—curtsying and displaying their angles around a gymnastic pole—expose their parts with piqued indifference or indolent pride. The men pay for more intimate settings in which the object is held just above their faces to stare into, dangled teasingly in front of them, or actually pressed into them. They sit in numb, self-seduced languors, overstimulated with nothing for their minds and senses to go to except a language string, bottomless and self-refracting. And why? What are they looking for? What are the women purporting to show?

Subtlety has been scalded from the world, while a vicious and implacable literalism drives us out of the profundity of our senses toward imagined ecstasies and consolations. A kama sutra of epiphanies and delights is subverted to a purgatory of disembodied acts and viscera. This is not passion but its opposite.

Hyperconcreteness rules the day, a desperation to get to the bottom of, to possess things that attract and shimmy, by going at them rather than the latency behind them, to divulge the cherished object more and more vilely and explicitly, as if raw anatomy and sheer size—labia, clitoris, tumescent penis—contained the feelings, the riddle they elicit, instead of vestiges of clams or lizards they also once were.

The customers can watch and probe all they want, and they can even stage performances, but they cannot touch and, even when they do, they cannot reach the object of desire. Gazing at lobes and orifices of vaginas, mesmerized by the facsimile of allure, they do not see the event any more than astronomers espy the source of the universe in more and more distant galaxies, greater black holes giving back dubious information, explosions implicating trillions of stars. These cataclysms are so far beyond imagination or calculation they do not enter the equation.

Our obsession with smashing the atom, with capital beyond equity, with pleasure devoid of experience leads to a pornography that keeps tearing the veil and releasing energy without satisfaction or resolution.

What could the bottom line be? People exhaust their lives accumulating the symbol rather than the thing. They pack heat, race trucks and jet skis, hoard dead presidents, pay to shoot exotic animals on ranches. "Hit and run" has become epidemic, particularly on the boulevards of cities, even children crossing the street to meet the ice-cream man. Money accumulates without real plenitude. Tax-sheltering it or converting it into stuff is tedious. If you buy everything and deprive everyone else in a zero-sum game: all you can do is win, win, win joylessly,* until you lose, or until the rules change, or there is no more game.

To intimidate "truants," cut off their limbs or their ears. Or be done with it: machete them in the streets or machine-gun them in courtyards and stadiums and pile their corpses to be gaped at. To enlist them as customers, innuendoize sex and then con them with sequins and foils. The mission of lust, like that of dominion, is to keep pushing the envelope and bullying toward an absolute denomination, an ultimate terror at the rad, *noir* heart of matter.

Paramours behave more lustfully than their actual desires,

---

*Ask Enron's late West Coast energy traders or the Steinbrenner-era New York Yankees.

gourmands more greedily than they have appetite, soldiers more cruelly than they have emotional capacity for. People drill counterfeit grimness, ferocity, and casual hipness in stases of puritanical rigor. The walls of the modern museum are hung with canvasses replicating this gaze. Constabularies expose, pound, humiliate, liquidate, and goad into submission because nothing less will do. As Chemical Ali and the perpetrators of the Hutu *interahamwe* taunted, "Who is going to stop us?" When pimps (and worse) descended on South Asia to round up orphans for sexual slavery after the 2004 tsunami, the argument of the world approached its own abjurement. Who *is* going to stop genocide or the ravaging of the planet and extinction of species if we can't protect these kids, if we won't pull over at the scene of the accident, if we can't restrain two-bit thugs or even our own disengaged ennui, if all attempts to deter are merely collaborations at another level? How do we expect to hold off terrorists when we inflict so much terror ourselves?

Our culture now rests on artificial and pedagogic restraints alone keeping us from the amoral max.

Maybe it has always been this way—piracy and plunder, the spoils of victory with excusable collateral damage—maybe we are only now turning to face ourselves.

"Who is going to stop us?" like "Fuck the international community!"—a battle cry that officials on all sides, tyrants and professed freedom-lovers, appropriate—a dare to God to step in and end this carnage, to seal the sky, to make it safe and even valiant not to have to plunder and desecrate any longer.

They may not know it, but bullies, rapists, and torturers are pleading for someone credible to tell them there *is* a limit, to assure them that the limit means something (that they won't be cheated of their spoils or honor or occasion for release), to give them a convincing excuse to halt their ruthlessness, to liberate themselves. Until then, the world will be led by marauders and torpid hedonists who imagine limitless danger unless they prove themselves truculent, pitiless, and immune, who fear that the enemy they leave standing, the object of desire they spare, will be their final undoing.

The men want, the men need, but, instead of going toward, they pry and search everywhere else. Instead of transforming ceaselessly what is inside them into the unexplored danger of this amazing ecstatic life, instead of entering the prismatic, formlessly forming universe that gives birth to every new act, they remain riveted on the snare, holding it compulsively in place, to possess the experience, to grab what is there before they are not—and nothing changes, and nothing frees them or the rest of us held in their thrall.

The fact is: everything is real, and anything, once the spell is broken, can rise wingèd into the sky.

With relentless certainty, giant waves toss and spin this small body of mine as casually as they hurl a Kraft Cheese wrapping and the bottom of an old barrel in the increasingly littered North Atlantic. I taste salt and childhood, an earlier planet. I stand up dizzy in the surf. In the haze a low-flying plane appears, then presto—gone!

An Alternative Explanation

It's not that one doesn't know he or she is dreaming. A person always knows that "it's a dream." But they also always know that knowing changes nothing, that the dream has to be dreamed for other reasons, so they go on dreaming as if it were real. Dreams, Freud said, guard the gates of sleep.

A fearsome scenario in comic books of the 1950s involved insectoid invaders from another solar system, who conquer the Earth with their superior weaponry while pillaging and slaughtering without pity or avenue of appeal. They place their leaders in the White House, the Kremlin, and other centers of power, and from there broadcast genocidal strategies. After all, they are the powerful sentient species; we are the bestial underlings—why should they have empathy for us?

Yet these cerebral cockroaches tend to say the right sorts of things: "We bring higher intelligence to save the Earth, to ensure a bright future for both species." Everyone knows they are lying—much in the way the invaders of the movie *Mars Attacks!* dissembled (to cheers): "We come in peace!" That was moments before turning their ray guns on the concourse and skeletonizing Congress. The words, needless to say, were empty—diplomatic drivel, crowd control. The Nazis could be very reassuring too, as they coaxed people into railroad cars and gas chambers. Even the 9/11 hijackers told the passengers about to be flown into the World Trade Center that they were returning to the airport—please stay calm.

In the same way the Bush Administration is terrifying. Morally conceited, lacking compassion as bellicose insects do, the President and his colleagues pretend to hear the pleas of the suffering; they use a quotient of right words: "Bipartisanship, freedom, our allies (accent on the second syllable), no child left behind, clean skies, tax relief, democracy." The ominous sense is of smiling, antipathetic mantises who have no feeling for any of us and would exterminate those who oppose them in a heartbeat.

Magic without science is stagnant and terrifying and puts humanity in a nostalgic spell: anything can happen; anything can change into anything else—and did for millennia. Witches and conjurors ruled the streets. Those who control "magic" control the "trance" and so fashion nightmare after nightmare, sacrifice upon sacrifice, witchcraft trial after witchcraft trial, evil eye for evil eye, only a fraction of them real, the rest projections and scams, gauntlets and reprisals. Thus, finally even magic and its transformative power are lost.

Enter the reigning gestalt of science everlasting. As the electron microscope is shined on the "real" causes of phenomena, magic is unmasked as flimflam, spurious agency; its practitioners are relocated on the fringes of civilization. But science without magic is sterile. It delivers a universe of robots and orcs. In a scientocracy humanity doesn't participate, merely enacts a travesty of "being"—a pragmatic demonstration of what it might be like if things were real. Existence becomes artificial, a series of considerations, a momentary illusion of having won a cosmic lottery. And then, at the end of it, snuffo! —a savage beast consumes body and mind and any memory of "being" as though nothing ever really happened or mattered.

Fierceness avoided is what is left to confront, after the big tent closes down and morning no longer follows night, in alien lands ruled by death.

But science is necessary for progress, especially when it sponsors experimental empiricism over materialist ideology or academic makeovers. The unbiased human mind probing objectively into the matter in which it is clothed, asking the single unwavering question—What are you and what am I?—is different from science serving rank or greed.

Magic after science has some hope of recovering the core of our nature. But unless it is truly post-scientific, we will be catapulted back to the Middle Ages or voodooland; we will trade reason for superstition. If (on the other hand) we run enough reproducible experiments deep enough into nature to embarrass and topple the corrupt priests and conjurors, the fake astrologers, the jealous witches, the exorcists, if we can keep them from regaining their stranglehold on society, then we can begin to claim, one by one, the real treasures of our psyche, the operating energies, the premonitions, and quiet epiphanies by which we may (in the right circumstances) transform even cells, even molecules, and heal the troubled waters of Babylon.

To practice universal field medicine and compassionate breathing in place of HMOs and corporate pharmacy; to read cards and stars and sacred alphabets without losing the thermodynamic paradigm or forgetting how to run the machine; to restore the soil and water biodynamically and alchemically, by transmutations and infusions we can't begin to parse, are the only hope of salvation in a scientistic time, are the only reprieve for our shimmering world, the only path to deepening the texture of existence, becoming whole and complete and individualized in a universe which was and is yet our birthright.

Bill Kotzwinkle described to me the most amazing martial move he ever saw. A young alligator, a few feet long, was lying dormant. Bill approached it from behind out of curiosity because something didn't seem quite right. Without warning, the animal reversed itself so that its head was instantaneously where its tail had been, and it had levitated, barking right in his face. "What a use of chi!" Bill exclaimed.

It had been practicing that technique for the last five million years.

In Anita Shreve's novel *Light on Snow*, Nicky, the twelve-year-old narrator, recalls losing her mother and baby sister in a car crash two years earlier. The family's fairly normal life had been taking place in suburban New York City but, as soon as feasible after the accident, her father oversees packing all their belongings in boxes, and then they head north in a car and trailer in search of a new home without memories. Doubling back off the freeway into an anonymous southwest New Hampshire town, they pull alongside a real-estate office. In the front seat Nicky has contracted into a stubborn, angry ball, as her father wearily attempts something like a pep talk, alluding to future plans as if this were all normal.

Suddenly Nicky feels frantic and disoriented. "Where are they?" she cries out.

"Where are who?" her stunned father replies.

"You know who. Mom. And Clara. Where are they?"

Defeated by this demand, he pleads with his eyes for exemption, but she cannot let go of a bottomless rage. "I hate you!" she screams.*

Though the human heart, not cosmology, is Ms. Shreve's goal, the only other answer to this question—other, that is, than rage—is another question: "Where are *we?*"

Somewhere I read this koan: if you were offered a hundred million dollars for your life, would you take it? Most people would rather live. But it's that moment's pause....

Memory has no leverage against the vastness of the cosmos or the chronicles of eternity. Whole worlds plummet into specks of dust and then nothing. The universe sets no memorials and keeps no records; creation is its own testimony.

Why should mind not experience complete amnesia of its life after it is over and a new phase begun? There is no membranous sac between existences in which to transport memories, no matrix in which to document loved ones and events, no transitional vistas either. The figures of each lifetime melt away like figments in a dream.

Bio-memory is a glass vial for creatures in bodies, a fleet-

---

*Anita Shreve, *Light on Snow* (New York: Little Brown & Company, 2004), pp. 136–37.

ing translucence bearing a semi-stable clock. When we take on bodies for documenting episodes and events. When we take on bodies, we cast sticky ganglia, gossamer threads much like the silken geometry that exudes from a spider or the web of slime discharged by an eel unintentionally landed on a fisherman's trawl. This gelatinous artifact synapses with dews of illusory landscapes and stores episodes we paladins consider real. Not true memory, it provides static context for images, isochronic maps for survival. Take away the context and the images and plans evaporate, always. Without context we could not remember from moment to moment what we just thought, let alone across the mortal veil.

Compared to timeless memory molecularly rendered in moraine and sky, biological memory is an evanescent hologram. When we shed bodies, we join the single identityless unknowing. Once the incarnation is gone, memory is gone too—total Alzheimer's—to be replaced by nonmemory in which every wisp and particle, every dust bunny, every stone, fissure, and iota of the actual universe is recorded forever. Only nonmemory remembers every encounter and everyone that ever happened, arranges abrupt yet credible awakenings of creatures, equipped, embodied and minded, into worlds. Only nonmemory thinks without distractions. Only nonmemory reads the hieroglyphics of long-extinct worlds, projects radiance into matter, karma into form. Only nonmemory archives this moment forever.

Listening to the inimitably sweet lilt of teenage Sam Cooke on the re-release of the Soul Stirrers' 1951 gospel, jamming with his elders, an avuncular quartet, just about reinventing them with his silken croon—in particular "I'm on the Firing Line" and "Christ Is All" *("there are some folks who look and long/ for this world's riches, oh yeah ...")*—casts this haunting nimbus of innocence, of African coming-of-age sincerity over a career yet to be, from the equally soul-stirring but downtown "Chain Gang," "Cupid," "Another Saturday Night," and "Wonderful World"—the "Best of" soundtrack of a decade of romance and loss in America, a whole anti-Sinatra universe headquartered in 1957—to the club balladeer spinning out elusive dirges and enchantments of "Jamaica Farewell," "Bali Ha'i," and "They Call the Wind Mariah," to the orgiastically pounding R&B, pre-rap Harlem Square revival, 1963, "Bring It On Home to Me" and "Nothing Can Change This Love of Mine," a year before he was shot dead at the Hacienda Motel, $3 a night, L.A., chasing a prostitute into the manager's office—an event which, for some, like the Kennedy and King assassinations and satanic gunning down of John Lennon on a New York sidewalk *("... don't you know that he's all,/he's everything to me ...")*, was the beginning of the end of the party even before it began.

Consciousness is a primal commodity like gravity, heat, or charm. Our awareness of our own existence is universal, not personal, much as electricity flowing along power lines in Nepal is the same as lightning on a planet in Andromeda.

When I was younger, I considered the notion, advanced by Hindus, esoteric philosophers, and psilocybin devotees, that this life is but one of many planes of reality—equally valid dimensions, each packed with solid-seeming objects and beings, yet actually vibrations manifesting at different pitches. I believe it was Ram Dass who said that dying is like taking a trip somewhere else, as if one were to go to Australia, nothing more. Timothy Leary compared separate worlds to channels on a television that can be switched by turning a dial that changes the frequency of the waves making them up.

This always seemed eminently reasonable, if a bit facile. While believing it, I thought that this life, this "me" was real, an actual moment in time, the world the realest world because its narrative rang true and it felt uniquely, grimly serious. Like most of the population, at some level I bought into the proposition that life is the whole shebang, crucial in an irrefutable way.

It now seems that, yes, we are condensed oscillations cinematizing a virtual landscape. Stuff is porous and electromagnetic, and life is not a thing as much as a probationary ripple. Like everything else big or small, we are composed of fumes, phantoms even when colliding. The flesh may weigh in stones, but stones, my friend, are gathered light. The rough-hewn carcasses we carry around answer but to slave names—they are not it or us.

Our existence and that of other creatures is radiation tuned to the frequency of solidified ectoplasm, a bundling of substanceless threads with ephemeral charges into body/minds. Some greater magic infuses vapors with consciousness such that scenery from this vantage is palpable, stable, and compelling, and our personal fiction grounds itself in the hallucination of its own physicality.

So are we plunged into the imaginary flow of time, a stick into a stream, no way out except where the stream is going, into the future. It little matters whether time is real, because we are imprisoned inside it. It little matters if bodies are real because we are tuned to their precise pitch and resonate materially with them, and consciousness apprehends its accoutrements as identity, absolute and self-existent.

The entire universe in fact is an emanation of smokes, harmonically kinked along its band-width, burnishing subatomic particles differentially into events (even humongous celestial objects are soft eddies that can be crushed to thimble size). It is not "being"—wherever, under whatever circumstances—which is solid; it is knowing, the way that being is perceived, that makes an energetic pulsation valid, a tone into a world.

That's why Buddhists call this (and anywhere else) a bardo. Artifacts or epiphenomenal semblances, we serve an unclaimed hypostasis, puppet extensions of our own psychic fact.

Put a body under the highest magnification, and there is nothing there, electrons orbiting nuclei inside whose fields warrants of even finer energies fluctuate between signs and gaps. We are ghosts that do not look like ghosts and, more critically, do not feel like ghosts from within. To ourselves and our plans, in our cities with our machines, we feel like the antithesis of "ghost," a proprioception that all self-important ghosts have, that their situation is substantial.

Giving up the ghost is exactly that. Existence seething with information and urgency is whooshed away, dispersed like steam. Nothing—not a sigil—remains.

None of this makes life false or unimportant. It is probably as real, or not, as anything.

We are matter doing time.

Existence has hemorrhaged out of the stark, empty profundity of nonexistence. But reality is not a vague, malignant bulge, splattering lava and dust throughout universes; it is a concise, discrete patterning of intrinsic dispositions in molecularized sequences. Self-replicating threads of cell spatiality form explicit bridges between not-being and being. On Earth, strings of DNA dilate holographically into blastocysts and ignite the Breath of Life. This routine brings us, however we are—unnamed, unarmed, unidentified, and mostly unconscious—into an existential realm.

Elsewhere among vast multidimensional universes creatures may have more or less concrete or corporeal ways of incarnating themselves and coming into cognizance of existence.

The question: what is life? has untold numbers of partially satisfying answers: Life is the sole qualifier to a visionary experience, protracting into a body through pupae, manifesting on material worlds in sacred geometries. Life is a chord, an attenuation, of creation. From its vantage we see the rest of, if not the creation, if not the whole, then the universe consolidated and installed in our existence.

Life is particularly prized and unequivocal because it glimpses not only itself but the scenery from within to without in a form so voluptuous and stark, so riveting and enchanting, that it would seem nothing else is real. Certainly nothing else is alive in this way. Life is consummate and irreplaceable, though not unconditional. Road trips like these come along once, in a lifetime.

Isness can never be as nuanced and deep as is-not because is-not incorporates everything that "is" as well. "Is" is the multiple, myriad, diverging shape of nature, while is-not, in its undisturbed potential, is everything that might and will be, that was and is. An egg is always *rien.*

Nonexistence is not obliteration, nothingness. It is *partout,* realized or unrealized (however you will have it), of which these lives are partial manifestations.

The advice of the Buddha was to dwell in the wonderment and swift but passing phases of "what is," the ever-vanishing present enchantment, never losing track of nonexistence, of the vast, empty sky both without and within, the exquisite subtlety that supports and sustains us.

It is the profundity of nonexistence that casts existence in a ceaseless intimation of profundity, of having fulfillment and completion slip ever through its grasp. Existence knows how profound it really is but is not able to locate it in actuality. It appreciates its own paradox and glimmer but cannot pin it down to any moment or thing. Instead, it does the seemingly impossible—make more of eternity by fracturing it into the responsibility of time.

A joke based on whether "one" is a noun or adjective:
   What did the hot-dog vendor say to the guru?
   "Make me one with everything." *

---

*I initially thought that my craniosacral therapist dreamed this one up. But these days it's even on bathroom walls. In another version the guru also hands the vendor a twenty-dollar bill. The vendor pockets it, saying, "Change comes from within."

There are countless universes (multiverses), each with its own principle of formation, its own laws (no gravity in some, no light or matter in others). This idle proposition from a cosmological think tank applies an impeccable algorithm to that mysterious self-knowing called "I." If there are myriad (like a trillion) multiverses, only a few—perhaps only one—will awaken, so we hypothetically exist in that single universe uncommonly equipped and predisposed for life forms like us, while the rest remain *au naturel,* entropic, higgledy-piggledies, most without hope of a dust mote, let alone a cell. This preserves accidental etiology and spares biologists having to deal with the anomaly of consciousness.

Even in this singular orderly universe, among its trillions of worlds, only a handful are conducive to life. On our exceptionally syntropic planet, millions of differential Darwinian years still must elapse before Camelot flickers for its brief shining moment, between ice ages, between nitrogen and greenhouse atmospheres, in which we are present, because where else could we be?

Thus is spiritual nihilism maintained, the algorithm pushed to satiety. Thus has spirit been wrested from us yet again by clever skeptics and sheer didactic force.

We humans no longer have even the self-confidence of a newt or a beetle.

On the other hand, there is a preexistent, timeless self-knowing ceded life sentences of indeterminate numbers of days and nights in shapes without parole. Thus is the dharma established.

Creatures carry out what their situations entail. Each of

them is a cut into a fabric of multiverses or what have you; i.e., each is too complex for the algorithm, except insofar as the algorithm is generated out of their own problematic and troubled self-knowing.

There are only views as such—the rest is void. No mountains, no stars, no gravity or light.

Three a.m., lying in bed in a state of restless insomnia, I am beset by urgent, obsessive thoughts. There is no way around them. They arise from disturbing events of the day and, even as I fall asleep for a moment, startle me back up. An accepting, hopeful outlook, that allows me to glide most nights into slumber because this is the world and I am here and tired, no longer functions. Day and night have fallen out of their simple cadence. Nothing soothes; nothing works that gentle transition into scrabble. From what little distance I can gain internally, I search for the root, but all paths toward it are in agitation. I stare at my erratic thoughts, follow their trajectories; they are unstable, singsong, jagged, desperate, grasping for solace or meager amusement—I plead, "anything but this."

I make a crisis decision, to stop trying to distract myself—no more elaborate fantasies solving international crises, no more toppling regimes or conducting "Missions: Impossible" while invisible, no more landing onto alien landscapes of other planets, no more women and narrative seductions, no garbage of any sort to trip myself into sleep. These are all only postponing the inevitable—confronting bare mind as the source of disturbance.

The object of my worry is not really the thing that happened. That may have been the trigger but, powerful and upsetting though it is, it doesn't mean goose down against the greater darkness in which my troubled thought contends for its nature. The real basis of this anxiety is that my mind has lost its innate assurance and bearings; it has no place to roost, no fundament to attach itself to, nothing to count on, no way to abide or be comfortable, no confidence or interest in its own situation. It has been consoling itself for a long time with merely one distraction and rationalization after another, called adult life, until something too threatening and insistent to permit distraction occurred. Now it races around like a shade, anchorless ... settling, alighting, settling, alighting.

In childhood my brother used to leave his bed in lost watches of the night and stand in our room punching at air, grunting. I would stare at him from my pillow until he noticed. "I'm fighting ghosts," he would tell me in a harsh, impatient whisper. I accepted this uncritically, but without understanding the actual feeling in him or why he called "it" ghosts, what compulsion led him to appease it by thrashing at emptiness. Decades later he told me he would call up his worst terrors and try to pummel them out of existence.

However we act it out, ghosts are what we are fighting. The Dzogchen teachings say to welcome and engage them. Give them space. Go toward fear always, as a supplicant, not with the illusion that willful courage or battle is what is called for. Don't crowd anxious and unpleasant thoughts. Create room for them to roam and expand.

Tell the brain that it has been working too hard and now it should get some rest. Tell the other organs, one by one by one: heart, liver, kidneys, lungs, spleen. Acknowledge their existences and steadfastness and grant them dispensations, the kind which mind alone can extend to the precincts of its body.

Relax the facial and eye muscles, their habitual patterns and potential toward patterning. Relax all residual expression and the tendency to expression. Do not keep pecking at aggravation; it cannot be subdued or undone. It must be softened, gradually, into grief. Do not placate or pretend to minimize stirrings of panic: that will enhance, not dissolve them. Do not get frustrated, and do not take hasty flight. You can't think your way out of this. You just have to live it until breath changes it into something else, until experience becomes tolerable again.

Relax into unconditional body and its million-million-cell circuits, free of life, free of death. Just keep coming back to the tangency of this and that, whatever they are, and the playfulness and innocent wonder of it all.

Adversity is our only opportunity to practice, however half-assed our practice may be. It is not about perfect practice anyway. It is about mind having its own intrinsic nature, there all the time. However sincerely you concede even the smallest inkling of this, even briefly, you find a temporary roost.

At death the conditioned mind, shocked by sudden dissociation of a stream of input through the senses from an external world, imagines it is nonexistent and faints. But, if mind is dead, what is imagining something no longer anything? And what is death—or life—if there is only a state of mindedness generating phenomena and projecting itself onto worlds?

Eventually the mind wakes where it is and realizes it could not be dead; in fact, its basic nature is much as it was in life. But now there are no words to propitiate it, the world about it is unlit, it cannot touch or be touched. At most, it can throw itself against solid objects and grab at living bodies; but the contact is fugitive, the collision empty, opaque. With tremendous effort it can topple a vase or grab a leg, startling an embodied person; abutment fizzles instantly: a specter cannot hold matter. Beyond caress or physical solace, bodiless and a spirit, it still perceives and is subject to karma. Preferring the comfort of solidity and arousal, of its own remorseless projection into space, accustomed to being enclosed in the shawl of its recent body, it repropels the cocoon we experience as incarnation. (It could twist itself in any number of states or shapes, and they might become worlds like this one, creatures of various sorts, some quite weird by human standards.)

I don't know how embodiment happens, but spirit mind, exigent in its plight, scurries to get itself born in another body, a frog mechanically returning to its pond.

I remember indeterminate moments of my early childhood, probably around age two, lying on my back in a crib and gawking at the ceiling and drapes, grabbing under my quilt. My legs kicked awkwardly, as I was suspended in the spaciousness of it all, the simple awe of being, the wonder that now a point in time is here, at me, in this room, solid again and real.

I began to explore my body with wonder and delight—bending finger by finger, feeling toes and the spaces between them, chasing slippery testicles, putting a finger in one nostril and another, around the ear canal and snug into its hole.

I cannot help feeling I knew something then implicitly I no longer know.

People behave as though this is real, a special place in a valid universe: the magnificent streets and shops of great cities; pedestrians in all manner of regalia and personification; forests packed with quadrupeds and birds, equally convinced of their condition; the very ground infested with microbial, bug, and worm life that has no consciousness to speak of, except the eensy bit it has, that exists to exist. The unabashed packedness and translucence of space is such that everywhere it extends, in all directions in relief and scale, it has something in it, plants towering out of its underbelly, men and women congregating to its horizons, cars in endless streams on its pavements, interminable seas, stars sprinkled over black time, wasps swarming upon fields of blossoms.

This reality goes on for staggering distances and then ceases by not ceasing. Beyond earth and sky, beyond the universe, is simply more space, inside itself and out, on and on, forever. Beyond mind and rigid ego-identity is the chasm of death, the place of dreams and shadows, generations reeling through the dormer, history spilling across a beam of omniscient daylight. Everywhere the stage extends and stuff is happening, something is curtsying and paying homage to the real. Even the trunk of the apple tree is stuffed with giant pellucid beetles about to be born.

What we have, though, is a bubble: an incredible collective hallucination, a consensus reality that, however brave and impregnable, everyone is working desperately and overtime to maintain. "Find the pivot point," Don Juan Matus confided to Carlos—"Strike through the mask," Ahab told Starbuck— and you will uncloak the truth: the fiendish, malign entity that no one wants to see, that everyone is collaborating to keep under wraps, the blank, terrifying real.

What a show! What extraordinary effort every second to hold the lid on, people hypnotized and sleepwalking 24/7, just being themselves.

We are the ultimate hand of cosmic poker, the billiard ball coming back after hitting the far rail of creation, knocking #15, red, the object ball, into the corner pocket.

We have all agreed to be here and play by the rules or, stated differently, anyone who didn't want to play by the rules is not here.

My daughter is having a remarkable run with her first film. After winning a number of awards in the U.S., she has taken it to Cannes where it is being praised as the American sleeper hit of the year. She wrote and directed it and co-stars in it with John Hawkes. It's called *Me and You and Everyone We Know.*

In a snapshot-and-journal blog,* she talks about her trip to Europe. She says "she could have made a documentary about a young woman who goes to Cannes with her first movie.... She wrestles with feeling frivolous for enjoying all the attention and pretty makeup and dressing up. But in the end she decides that it doesn't do anyone any good to feel guilty about it. Guilt won't reverse the effects of global warming, not even a little bit. And who is this god who decides what is frivolous? This god is nowhere, she decides." Or maybe a vestige of some childhood memory. Later it "was decided that Dior would loan me a dress. As it turns out no one actually owns those crazy dresses, but maybe everyone already knows this. Here I am in

*www.meandyoumovie.com

this dress and feather cape, marveling, overanalyzing what it means to wear such a dress, to live such a life, as the water levels rise."

It is amazing to have a wise daughter, and reassuring given the planet she was born into. She is not fooled by Hollywood glam or her anointment as iconic indie-artist cover girl. She will watch the tragic procession across her time attentively, soberly, courageously. She will transgress societal boundaries always, but respectfully, without prurience, without contrivance of agenda or ambition to shock or betray, as she maps a world of lonely people, young and old, and their beautiful chatter and follies with a wide-eyed appreciation of it all, the strangeness and yet ordinariness of lives.

That tiny blue-eyed thistle performing bath-tub theater matured into an adolescent protecting dying birds, reassuring them about their crossing. She mutated from a blonde psychic dreamer with stigmata on her feet to a dark Mohawk-crowned 'zine publisher who directed and acted in her own staged productions at the Gilman Street punk club. After dropping out of UC Santa Cruz, the playwright of *Sink You Up* and *Fingers and Liars* became the narrator in "riot grrl"-bands out of Portland, Oregon; then of edgier vignettes in nightclubs, solo.

A young woman standing on a chair, she shot out both arms to deconstruct the phrase "star" before she could become one ("movie star or shooting star?" she asks). Without solipsism she impersonated her own sight-threatening corneal disease and the "alien abduction" ophthalmologists who examined and interrogated and cut holes in her (such that seeing and "being seen" morphed into each other, and the barrier between sexual and scientific broke down, between violation and treatment, as medical office replicated fantasy booth and to seduce and examine became different ploys of the same impassive authority). She converted drunk and recreationally obnoxious hecklers, reinventing their jeers for them on the spot, always honest, always fair.

From the chrysalis of a girl with a mike emerged the multimedia performance artist of Lincoln Center, Rotterdam, Liverpool, Vienna, the uncannily brilliant director of children (Berkeley's most creative babysitter and one-time grade-school impresario and play-date emcee, having become what she always was), the story-teller of the Whitney Biennial, who discovered the "independent crime-crime-crime-crime-crime ... the kind you have to say five times ... and you can't be sure it really happened or if it is really criminal." She travels to the scene of it on a small train: the engine is the word "independent"; each of the elusive crimes is a boxcar. She finds "sheets ... pulled over bodies," not animal or human, just dead, "the shapes of things that have been lost violently." She sees "the terrible ferns ... that only sprout at the scene of an ICCCCC."

Miranda is a person to whom I can tell my problems and get back answers that are at once prophetic and unflattering, that reveal how to take the next step, even if it is impossible, and the way to make it possible.

She is profound, unrelenting, full of heart and eagerness. They can't slip a fastball by her, and they can't fool her with a curve. I only hope she lives in a world as big as her vision, one that will allow her gift its mortal expression, so she can return a finished work to the gods.

"No one else will ever come here, and that is the worst part, to be the only one who has seen it."

Ten thousand years ago it was "Stone Age" here. Men and women were still in an animal phase, managing the hunt, burying mummies of numinous cats, tincturing plants into talismans. We have driven automobiles and trains out of forests in which there wasn't even the hint of a wheel, let alone an engine, and we have strung information across the clouds. The transmogrification that has taken place on this one world in a swift ten thousand years can only be a concrescence of something hermetic and essential in nature.

One thousand years ago it would have been impossible to foresee urban civilization with its smelters, newspapers, paved concourses, and telecommunications; yet it is harder now to conceptualize five *hundred* years into the future.

We are at the end of something. Despite the primitive, raw landscape of the classical late Stone Age, ten thousand years ago we were already well on our way to completing this entire phase of existence (read Parmenides and Anaximander for a mileage check). We can still glimpse (more or less) the forest of symbols of Middle Earth from which we have originated. What we cannot imagine is what will follow the coma of techno-science, the trance of conditional awareness. When we hear the call of Pan again, from what quarter will it come? What will men and women do? What will occupy Manhattan—a ruins or a new metropolis?

Take away this lifetime, and there is only a trans-millennial event, like the stirring of a gigantic anaconda.

The ratio of water to dry land on this planet was set long ago at about three to one, though it is under continuous revision. Ice is a third factor, aggregating sea into pale mountains and then giving it back cataclysmically. Even the largest tanker, and certainly this ferry, is tossed by gigantic waves that crash against its vector, its flotation chassis of complicated driftwood and mind-infested gears.

The ferry is of the machine class of things and propels its mass against the incomparably greater and quieter will of the waves. With its passengers it comes into the vicinity of the island surrounded by sailing craft, two kayaks, and jetting dung beetles, also of the machine class, that spread a noxious aerosol while secreting tar that will contaminate ultimately even the abyss.

Each isle with its grasses, shrubs, sand bars, conifers, wind chimes, beetles, ants, rivulets, Queen Anne's lace and clover, gulls and crows on promontories, is the Earth. The gate to the planet, the medicine wheel, is everywhere, circumscribed by water.

Paranoids are right about many things and wrong about everything.

A rose is made less of molecules than of time—time and emptiness. A carrot likewise. Also this purple stone. Time gathers into an exquisite moment, like a thought, and then exhales. How long a carrot or vein of pyrite exists is solely its breath.

My half-brother Jon has immense problems just living day to day and, whenever he goes completely over the edge, he turns himself into the local emergency room.* He has no health insurance, so the outcome each time is a multi-thousand-dollar bill against the dwindling funds that his father left, sort of, in trust for him (see below) as well as a prescription for medication. He doesn't understand that there is no "emergency room" anymore, just as there are (in the brave new HMO world) no doctors. He doesn't know that attorneys have recently defeated physicians in a high-stakes culture war (and this will last at least a generation).

On another level he doesn't get it that "his" lawyer is not the guardian Jaggers out of Dickens or even necessarily benign. He is stunned to be told one day by the man he took as his father's representative on Earth: "You have no trust; you're a pauper, a ward of the State. You just get the interest while you're alive. Your father left his money to your cousins because you're an addict, a lazy bum, and a bad Jew who never went to Israel or worked on a kibbutz."

What "Jaggers" doesn't say is that my stepfather's sisters and his last girlfriend got him off his death bed into a taxi, up to the office of their friendly family counselor in order for him

---

*Most of my writing about my half-brother and our families appears in two nonfiction novels: *New Moon* (Berkeley, California: Frog, Ltd.), 1996, and *Out of Babylon: Ghosts of Grossinger's* (Berkeley, California: Frog, Ltd.), 1997. Additional material appears in my essay "The Phenomenology of Panic," *Panic: Origins, Insight, and Treatment,* edited by Leonard J. Schmidt and Brooke Warner (Berkeley, California: North Atlantic Books), 2002, pp. 95–163.

to tape and then sign a new will and, according to my half-sister (who could have been played then by a combination of Dustin Hoffman in *Rain Man* and Talia Shire in the first *Rocky*), he wept over this hapless act for the last two weeks of his natural life, pleading with her to do something to reverse it. She assumed he was delusional. Now it is too late.

I know my brother's damaged. We all were damaged in that family, each of us in our own way. We were "deer in the headlights," roadkill very soon after birth. I live with the same demons in my head, but I've made a fragile homestead despite them.

My sister has done waitressing gigs the past thirty years, filling out lottery tickets and squeezing oranges for Manhattan flotsam, despite her Holyoke and Middlebury degrees and speaking mastery of three languages, her early career acting Molière off-Broadway. She hides in multiple costumes and personae, under the protection of carefully wrought superstitions, rituals, phobias, renting the same Hitchcock and Woody Allen movies dozens of times, eating only the same eight or ten basic dishes (a life of no milk shakes, no pizza, no spring rolls), and taking identical walks through New York City at the same hour to the same shops. She is like a Catholic, constantly making the Sign of the Cross against an unseen evil.

By contrast, my brother is in active counterphobic battle with sorcerers, witches, and even the yuppie dandies he was supposed to become. With his scruffy beard and dreadlocks and ripped clothes, never having pursued right livelihood (or any job at all beyond brief stints as delivery boy, juice blender, and night watchman), he wants to be recognized as an unflagging warrior against the status quo. Those who see him roaming through town call him "the Walking Man," as he shies from cars, throwing up a hand both to ward off their fumes and express his disapproval of their fact. Our sister hasn't seen him in ten years and won't even acknowledge his name ("Who? *Him?*"). His mere existence violates her profoundest taboo. He is a wildman, an embarrassment, and she is terrified of his possible return. She says she hopes he is found dead somewhere.

I think of our mother as a fiendish shamanic entity who tried to strike a single fatal blow on each of her three kids with some kind of baton. The sheer force and "kill" accuracy of that instrument are indisputable.

Demonic forces are willing to wait a long time to see their handiwork done. They disguise their weapons well. My mother's blow was called love.

My brother has been staggering for 57 years. He says he would like to jump off the bridge, but the water below is so dark he is afraid. Instead he is slowly bleeding, dehydrating, and scaring himself to death. He tells Jaggers he is going to kill himself. "Don't use a butterknife," the wry lawyer retorts. "Get something sharp. It will hurt less."

Though he refused to use telephones for decades in his boycott of the machine, Jon now calls me in desperation several times a week—and I reach him on my cell while driving between Berkeley and San Francisco. I try to counter his paranoia that his body no longer works, his pronouncements from the land of "no hope" and "too late." He won't be placated; he says he's wasted his entire life and is doomed: "I deluded myself. I thought I was a sorcerer fighting the establishment all these years. I imagined I was like Don Juan cultivating a mature relationship with a plant and its spirits. [Restless, uneven breathing....] That time you came to see me and I attacked you, after you left I thought, 'Good, I put him in his place!' I thought it was me, all of it, my joy in nature and my power, but it was the grass. I despised those Buddhist and Hindu fools for their fake Eastern wisdom. *They* were studying Advaita; *I* was reading Ernest Haycox. But they were so far ahead of me. How I loathe me! I was boastful and set myself above everyone. But at least they have lives. Now I'm too scared even to go outside."

I, the one our mother didn't love, was not whacked as catastrophically though, through all of childhood, the family damage was discernible *only* on me, gaudy and blatant in my acts of exotic anti-sociality, hysteria, wetting, and truancy. My brother did fight ghosts after midnight and sometimes, dropping his usual arrogance, pleaded for me, the official "bad" sibling, the one our mother called "evil incarnate," to intervene and pry his hand from the light bulb in his desk lamp because it was burning him and he couldn't break the compulsion to hold it there—but these were off-the-record acts that she either denied were happening or passed off as my snitching and jealousy when I tried to tell her there was trouble in our room.

Part of me kept panicking, vandalizing, running aground in novel ways; part of me stared from a place of absolute certainty and resolve. I meant to get out of that apartment alive, and I couldn't let them find that out—or risk knowing the true peril. I became a cartoon creature who, having walked off a branch, was doing stunts in midair without telling myself. If I had looked even once, I would have fallen. My best trick was preparing to be a normal adult by being a crazy child. I pretended the trauma and insanity around me were curable, in me too, by diligence or, failing that, symbols and fairy tales. It was effective at the time but bore an incalculable price. Now I spend half my life looking back in horror, amazed that anything still holds me together.

If we carried shattering events from haunted households inside us solely as bad memories, they would be far easier to amend and cure. But they spill formlessly inside our mentality where they corrode and mutate into other things, altered by fast-moving events of our lives and the immediate requirements of situations. They merge with us in insidious ways, hiding their connection to their source, spreading and changing, most of them dimly perceived (if at all) over time, reenacted as other things, things we imagine are novel and think we improvise on the spot. Indurate and intractable because we no longer know what they are, they become our acts now.* Body and mind experience their undifferentiated urgency, as if something tragic happened but we can't think of what.

The one gift my parents unintentionally gave me was permission to feel fear and to fail, in fact miserably. I learned surrender to forces beyond me; on my own I discovered the alchemical power of the wound, the value of aid from outsiders, the fundamental nature of redemption, a way to close off trashed rooms forever in order to survive. During those same years my brother, star athlete and scholar, was saddled with unattainably high standards, styles of bravado, wardrobes of success, loyalty to the regime and martial law, lessons in how

---

*This is why victims of parental cruelty often become cruel parents themselves, why so many sexually abused youngsters become pedophiles—to reenact and absolve what is buried inside them and give it voice, to try to understand what was done to them (and why it was done) from the other side of the act, to reclaim the lost and unknowable the only way it can be recovered, and to vindicate and, however imperfectly, redeem it.

to be a son-of-a-bitch businessman like his father's clients: the full cosmic gauntlet. While I goofed off or freaked out, he preened, looking on in contempt and disgust, hating me "like the devil himself," he now confesses. "You wouldn't even stand close to the plate in hardball because you were scared of getting hit, and I was ashamed of you."

That was before we fled the family story and learned to care for each other. That was before he fell apart in Colorado and came back to New York to throw our mother against a wall and show her his knife. She had him committed.

"It wasn't us," I tell him as he apologizes now for his arrogance and conceit. "It was them. They matched us like cocks in a fight. We never learned to be brothers who protected each other, who rooted for each other. I know you think it was a normal family, even privileged. Yes, we were sent to the best schools and camps. But it was a parody, a travesty of a family. We never learned what it was like to live in peace or to have a real mother. We had something else, a wraith who used a 'mother' camouflage. We never experienced her love, just the word, over and over till it had no meaning."

Three years later she leapt out the window ("a witch," Jon noted at the time: March, 1975), eleven stories down onto a courtyard behind Park Avenue. He was in the mental hospital by then, but her act freed him from the thorazine and stellazine being pumped daily into his veins, to three decades of wandering in a marijuana haze (until the ganja weed turned against him).

This I know: If our culture were not so corrupt and jaded, there would be less insanity, fewer addicts, less twelve-step and rehab, and we wouldn't walk a gauntlet of pimps, prostitutes, drug dealers, lost souls, and homeless. When one is already off-center inside, it doesn't help to step into a world that is listing more and more off-center. Even moguls on Wall Street know: this is crazy, and getting crazier. Jon smokes marijuana all day to take the edge off the dark vision and liberate himself into a land of salt marshes, forests, and birds.

Drugs soothe; they are the new authority, the communal mother. When all-night construction was initiated on the Old Post Road and marijuana no longer eased his insomnia and tormented dreams, he began to stalk aimlessly, pursued by spooks.

He is particularly oppressed by noise and fumes. SUVs look to him like metallic dinosaurs prowling the roadways of some apocalyptic city. A rebel against hi-tech culture, he pictures himself a throwback of the outlaw, an automobile abolitionist, as he strides the neighborhoods and writes papers on the frontier and American Western that no one will read. "Someday," he told me, "people will regard cars the way they think of slavery."

When he lopes down suburban thoroughfares of Westport, Connecticut, where fate deposited him fifteen years ago inelegantly and somewhat infortuitously like an urchin washed up from the stormy sea, he is attacked in body and spirit by the grinding whine of leaf-blowers, block after block of them. He says that it is like being in hell. If we were as sensitive as he, we would realize the accuracy of this observation.

Community is the sound of rakes on sod and stone, neighbors gently irrigating their internal organs and greeting one another. Madness is the destruction of community, not in single sweeping transformations, but gadget by gadget, event by event, gesture by gesture—bigger, more high-definition, more

TIVO televisions; *NY Times* fuck-you fashions; food additive by additive, Burger King by Burger King, guy and girl wise-crack after wisecrack, nihilistic pose after nihilistic pose, in lieu of empathy. This is modernism—the anti-hero equal to the hero, the cult of personality and loneliness triumphant over spirit.

He *is* in hell. He thinks ambulances in the street are following him, manned by wise samaritans who recognize his plight and mean to rescue and convey him to a sanctuary where he will be taken care of. Yet he is also afraid of them, so he turns abruptly down an alley or behind a bush and hides until they pass. Then he comes racing after, looking frantically every-where, ready at last to turn himself in.

Compassion flows almost blindly from the universe. It can-not be scheduled or assigned. His landlord, Ray, cares for him like an angel, even in the abysses of tortured nights, despite his own deficits and quandaries, with infinite patience and a categoryless love. He even invited his tenant to the local Kiwa-nis where the gentle rebel delivered a lecture on the history of the region from Indian cultures to settler raids and formation of the suburbs, including the origin itself of the Kiwanis lodge.

Miraculously my brother also found a wise therapist this month. It was the only time a certain Levy ever advertised in the newspaper. Jon saw the listing and left him a message. The call-back came two weeks later just as, in a state of unbear-able agitation, he was headed to the emergency room again. The trust fund quickly presented an obstacle, as treatment, even discounted to a third Levy's normal fee, obliged erosion of its principal; Jaggers wrote to the therapist, a document con-cluding: "In any event, Jon's source of income cannot support $450 a week on a continuing basis plus hospital admissions, plus his living expenses. That becomes a rich man's game and I do owe a fiduciary duty to the contingent beneficiaries of the trust."

The psychologist was quite direct with the attorney: "You know what it is for a bunch of relatives not to treat someone who's sick, to stand around waiting for him to die so that they can inherit his money—it's shameful, and it's a sin."

A week after I wrote this, my brother, perhaps emulating Indian warriors, perhaps taking the dare of his somnambulant ghosts, plunged a knife multiple times into his neck and abdomen, killing himself.

"I hope," said Levy, "the gods, or whatever there is, find a way to forgive him and grant him some peace."*

We have missed the real Earth, the planet here, for at least eight thousand years. We camp in a paltry corridor of reality from which we give little heed to the radiant colors of chakras, the ecstatic pulsations of all living things, the halos of mountains and ponds, the songs of invisible entities, the psychic waves undulating out of landscapes and forms, the third eyes crea-

---

*I sent this piece to my college friend Sid Schwab, a famous surgeon, who wrote back: "I'm not sure the gods need to forgive your brother. It's the other way around."

Joe Richardson, weighing in from Route 66 after his own decades-long journey solo on the wild side, said, "I don't want to get into any of the stock and trade phrases that can be found on Hallmark cards. It is my sincere hope that your brother gets a better deal, a fresh and new chance in the next world (whatever that realm may consist of)."

tures hold both open and closed, the auras that heal by touch and thought. In fact, the active organs to perceive them have become nigh dormant or vestigial in us.

When men and women perish, they disappear over a crack. No one hears their news; no one knows, for sure, where or if they go. No one glimpses the entry point of the living as they materialize over the transom. It is happening *inside* the real Earth. The dead go *in*.

In the Stone Ages, at least humanity had an inkling of the astral dimension of each being and the participation, even meddlesomeness, of ancestors in daily life. Australian Aboriginal Dreamtime is perhaps the brightest fading relic of Old Earth culture. It evokes not a preconscious or primitive scenery but a more complete realm, as when the lights are turned on in a parlor and its true size revealed. In the Dreamtime everything manifests both its becoming and its destiny, its evolving semblance in the landscape—everything is more than it seems, and the landscape is plenary such that creatures don't spring from nowhere or vanish from view forever. A waterhole that is a snake is a snake too. See that and you've got the whole game.

It is possible that fairies, elves, trolls, and goblins still dwell among great rocks and pools at different librations than us, some denser, some subtler. Tree sprites and cholitas emerge from xylem and bark by night or on rainy days and conduct missions and medicines beyond our ken. Sightings of invisible beings are quite common in Iceland, its geography of crystal, volcanic fire, unfinished creation. Children and constables spot them as they go about their business in their domain, using menhir-like formations for libraries, churches, and classrooms, proceeding along a frequency impenetrable by us but appropriate for them. Though tempted by beauty and sanctuary, few would enter a rock temple in pursuit of its returning residents for fear of getting shut forever in the stone.

In lakes of the north—Scotland, Norway, Labrador, and Siberia as well as Iceland—prehistoric "monsters" surface intermittently, not according to cycles of abyssal diving and breaching, but as the landscape re-tunes energy waves passing over the profound waters, causing vistas and creatures from millions of years ago (and times yet to be) to oscillate with the present loch.

The Dreamtime is what the Earth is actually like, and it is what we must return to after the industrial age and the fall of civilization, as we try to remake a world.

When there is every reason to be afraid, there is no reason to be afraid.

Months after our prior lunch at which we discussed "life as a computer simulation" my friend and I meet at Vanni's Restaurant where, as per plan, I ask Vanni to make us two Thai dishes using freeze-dried cane rather than the usual complement of refined sweetener. (A year earlier I had run into the chef at Berkeley Farmers' Market where he asked why I didn't come to his place anymore; I confessed that I had peeked in the kitchen and, while waving to him, seen Andy dumping sacks of sugar into the sauces and knew about the Thai regime of heroic sweetening. He promised that he would keep a container of organic cane juice on hand and would make me special dishes using it.)

After Vanni takes our order and promises a feast, my friend asks my thoughts about the present catastrophic species extinction. "Half of all living plants and animals will disappear in the next thirty-five years," he declaims. "It surpasses the impact of any asteroid or glacier; it is a cosmic cataclysm in a singularity, at least by the standards of geological time or human history. Don't you find it incredible that we're alive at just this microsecond when it's happening?"

I think about it for a moment: "No, because I don't have any conviction that time is real. Everything is happening all at once in some great cosmic mandala over which different modes of consciousness and identity graze. Time couldn't be real. Where is it going? Where might it be coming from? Everyone thinks they are living at *the moment*. Everyone is."

He counters: "But for most of history those are just little wars and famines. Right now it is the whole planet, and we're here in the blink of an eye to see it. That can't be an accident."

"It was always the whole planet," I say. "The real question is: are we the cause of this destruction or its agent? Are we in such big trouble because we are going through the labyrinth the wrong way, at least by comparison to creatures on other planets among the billions in the universe, or is this the crucible, the nigredo through which all consciousness has to pass in order to discover its nature?"

"Well," he exults, "if it's a cosmic lesson, doesn't that make it seem like a computer simulation? Isn't being inside an experiment a reasonable explanation for our present predicament?"

I agree: "Epistemologically the world has to be a simulation because it is all mind projection of some sort or other, from some consciousness or other. But that doesn't legitimize a science-fiction story about a master race and their computing superobject."

"What do you think?" declares Vanni, as he places before us asparagus and salmon, and tofu and vegetables (which my friend, as a pure vegetarian, directs to himself). I don't have the heart to ask if the salmon is wild or farm-raised.

Sometimes you just have to let one moment turn into another, no plans or turning points, no apocalypse and no pretext.

"Rice?"

Don Juan Matus informed us that the universe is a predatory beast, meaning not that it voraciously stalks and disembowels like a great bear but that it sends its fine percipient fibers through everything and pulls them tight. There is no escape from the universe's grip but, as a nonstop machine, its sole purpose is awareness and experience. Its every twitch and tug, each ruthless noose are innocent attempts to gain knowledge, to become worldly like us, to extend its own naïve omniscience into the city, into the realm of the eternal unconscious, into the deep and unknown shadows in which creatures conceal things even from themselves, into the darkness upon the face of the waters. We are creation's lovers and servants and scouts, its only hope. In fact, that is why we are here at all.

The universe is a predator of feeling and meaning. The only way it can expand who it is, can encounter the manifold unknown inside itself is by deploying us. This is both comforting and terrifying.

That bird seated on a distant wire in the misting rain is getting wet. No one can do anything about it; no one notices or cares. It was born somewhere in nature from the ovum of another unregistered bird. Now it is here, a luminous receptacle of everything, a portal to the universe. It burns with the same epiphany as every other creature, be it ant or midge, in a census or not. It is its own keeper. It knows what it is, not as we know it.

The eyes and the heart are cosmic and do not alter per bodily shape or location. This is true as well for those alighting ducks and the zoo-dwelling hippo keeping itself for as long as possible underwater, unaware that its asana can be seen through cleverly situated glass. It is even true for life forms without enough layers to make real sense organs or hearts. Once one of the trillion trillion eyes of Shiva opens, all mountains and seas cannot steal its vision or take away its prerogative.

The unconscious is not a padlocked closet or sealed papyrus within the brain or mind. It is the body itself, the tatterdemalion sack donned by the Breath of Life, congruent roughly to its esoteric shape. A coalescence of molecules flows from different nationalities of tissues, remitting nameless transmutations of stuff from lungs and intestines, atavistic personalities of glands and sphincters. These flood the mind's gestalt, organ by mysterious organ with alien minds of their own, enzyme by enzyme and in its whole, without explicit directive or statute from any cerebral center, without even being human, firing an inexpressible, unlanguaged void—some of it proprioceptive, some of it RNA background noise, some of it vestiges of ancient creatures, some of it cell and bacterial chitter, some of it happenstance rumble of neural traffic sucked upstream into cortical imagination—all unstructured sensations and images reckoned as mind. The body is a universe as deep, recursive, and barbarian as the night sky, imprinting as cryptic a cipher.

The unconscious is also the collective pre-atomic archetype invested in protoplasm. The amoeba becomes the preconscious worm becomes the unfathomable brain of the possum, containing the secret ledgers of all that is to be, all that has not been spoken. It holds battles, homing maps, libido, id, prime numbers, geometries, lacunae underlying future crises and metamorphoses, seeming memories of prior lives and other worlds, seeds of obscure languages, and vagabond impulses that arise from nowhere, have no explanation, and cannot be tabled or reinhumed, so must be lived.

Birds screech with the new day's sun; ants cross stone to reach tunnels of soil. Spiders, born for the short run, scribe secret texts, depositing bags of eggs, their spawn decorating trees with calligraphy. The depth of nature, including its billions of humans, its web of jungles, forests, seas, and their life forms, is totally unconscious, yet impenetrably rhetorical.

The unconscious does not erupt transpicuously out of chthonian zones into light. It gushes or seeps unremittingly, irrepressibly, because it has nowhere else to go.

The broken finch on the dirt has exuded a juice as bright (red) and hard as a cluster of berries, the mechanism under its feathers and cloak having leaked where it is torn at the breast. The

bird is a clock—the hour hand bent, the springs snapped—
that cannot be repaired.

John Holt is a singer whom few in America know. One part
Sam Cooke, another part Bob Marley, he has a bit of Smokey
Robinson too, a hint even of Sinatra. A legend in Jamaica since
the early 1960s, in 2000 he headlined a three-concert command performance with the Royal Philharmonic for Prince
Charles. With roots in ska/reggae by way of American rock,
his songs map a territory from Lee Perry's Upsetters and Peter
Tosh to Neil Diamond and Tom Jones. He has been accused
of selling out—glorified lounge singing—but also has been
lauded as one of the finest romantic minstrels on the planet.
His haunting subdued melodies could pass muster at a Nat
King Cole revival, but he also lays down powerful street mantras
in a hard, slow reggae beat. His CD photos range from the
dapper Brook Benton nightclub look of *The Tide is High* and
1000 Volts of Holt to the full rasta dreadlocked rootman of
Holt Like a Bolt (reefer out the corner of his mouth) and *Police
in Helicopter* (heaving a postal sack filled with something surely
illicit as a craft hovers just above him).

While his discs' liner notes are elliptical and not always con-
sistent from album to album, they are my sole source of information on "Reggae Mr. Holt" (as opposed to the progressive
educator of the same name), so I am borrowing their language,
reconciling them, and also interpreting them liberally to compose the following biography (which would probably draw a
smile from a Jamaican in the know).

John Holt was born in 1947 in Kingston. In a 1974 talk

with a British journalist he recalls, "School really wasn't my thing.... I preferred singing. I never attended singing class, though; I was scared. I was actually forced to sing in school by my friends. I didn't have the nerve y'know to really go out and do it."* Yet Holt won talent competitions as a child and at sixteen recorded the first of his forty-plus Jamaican chart-toppers. One of his rivals in the Kingston slams of that era was another promising young performer: "I used to whip Jimmy Cliff's ass, y'know ... he was afraid. If he knew I was gonna sing tonight for instance, he wouldn't turn up."†

Holt began his professional career with a fledgling producer, Leslie Kong, cutting a couple of records with Colsten Chen & the Vagabonds on the Beverley label, following with two more hits on Randy's Records, the latter while teaming with Alton Ellis whose prior partner, Eddy Perkins, had just relocated to the States. Despite these successes the Jamaican music business was corrupt and poorly funded, so a disheartened Holt dropped out for a year, returning in 1965 to team with Keith Anderson (aka Bob Andy), Garth "Tyrone" Evans, and Junior Menz (all future song-writers) in a group already known as the Binders. When Menz left to perform on his own, Howard Barrett replaced him and the guys changed their name. Recording on the Treasure Isle label, the Paragons, specializing in "one drop" Rocksteady, were immediately popular in the Kingston club scene. Their pre-reggae, Doo Wop sound was suggestive of many U.S. groups of the era—the Coasters, the Diamonds, the Platters. Once again, artistic success did not translate into

---

*John Holt, interviewed by Carl Gayle, quoted by Laurence Cane-Honeysett in liner notes for *John Holt: The Tide is High*, Trojan Records, New York, 2002.
†ibid.

royalties, and Holt and company kept switching labels, even launching their own Supertone imprint, a failure too.

Andy left the group early on. Barrett quit in 1969, departing for New York, followed by Evans in 1970 (after he had tried recording separately with Holt, then with Bruce Ruffin and the Shades—in the mid '70s Barrett and Evans were to revive the Paragons in America). Holt meanwhile, going mostly solo, put down a string of hits and was awarded "Best Male Vocalist" in Jamaica—twice, three times? He recorded with a number of producers, but his signature association was with Edward "Bunnie" Lee, who oversaw a smash success into the '80s. However, Holt's stature gradually declined when Bob Marley redefined reggae as indigenous politics and art on the edge. There wasn't much support thereafter for someone like Lou Rawls or Sam Cooke, willing to cover show tunes, country music, and disco.

Holt established himself in the U.K. as a club vocalist with a reggae beat, performing there first in 1968. During that visit he met manager/producer Tony Ashfield who later backed him on the Trojan label, the source of many of the recordings now available in the U.S. Under Ashfield's sponsorship Holt exploded at the Wembley Reggae Festival in 1970 and in 1973 began a decades-long stint atop the British reggae charts. He broke into the mainstream hit parade as well, putting his spin on American and Brit singers ranging from the afore-mentioned Benton and Smokey Robinson to Kris Kristofferson and Paul McCartney. In 1978 Holt made it to the U.S. where he performed with his old Paragon buddies, a collaboration that led to Blondie's cover of "The Tide is High" (see below) and a subsequent Jamaican revival of the Paragons with Holt, Evans, and Barrett in the '80s.

Despite an acrimonious breakup with Ashfield, Holt continued to work with Bunnie Lee in Jamaica. Though branded a "pop reggae" sellout, he surprised folks in 1982 by returning to his origin, recording *Police in Helicopter* with Henry "Junjo" Lawes, a reggae hit machine.

An eclectic artist, Holt is distinguished by alternately a rambunctious happy-go-lucky rhythm, hard slow beats, and sorrowful torch-song lilts with exquisitely soulful phrasing. The Holt songbook is varied and surprising over a prodigious range of lyrics and melodies, a number of which he wrote.

I discovered John Holt by accident in 2001 when I bought a *Best of Reggae* collection at Wal-Mart in Ellsworth, Maine, for something like $4.99. I expected to like most of the songs, but only three out of twenty hooked me enough to replay them. Holt had the third and twentieth items on the disc, the former a knockout. A few months later, back in Berkeley I went looking for something else by this "obscure" artist and found an unexpected bounty of foreign discs. Over the next two years I purchased eleven CDs at funky Telegraph Avenue stores like Rasputin's and Amoeba. Around my fifth a young male checkout clerk, groomed for punker tastes, gave me a thumbs-up and the accolade that opens this piece: "You're onto it. John Holt may be the greatest singer no one has heard of. The guy is awesome."

There are more than the eleven CDs I own, including some I can't find and a few I skipped because they have so much duplication that I would be buying a whole collection for one or two songs. An additional oddity: despite the frustrating repetition, neither of Holt's two songs on *The Best of Reggae* and only two of his most famous hits so-named in the liner notes are on *any* of them.

Here is my own top thirty from those CDs:

1. "Why Do You Hurt Me So"—Personal tastes vary from month to month and year to year but, if I have to decide on #1, it's probably still the original song that charmed me from *The Best of Reggae*. There is something eternal and haunting about its shifting inquiry up and down a mournful reggae drawl, from its first climax at *"Why did you make me feel so ashamed/in front of all my friends...?"* to its second at *"Why must you hurt me/and throw down my pride?/Baby, please tell me why."* The pauses between words are unexpected and stunning; for instance, after "all" and before "my" above. According to the notes, no one knows who wrote this song.

2. "I've Been Admiring You"—From the Bunnie Lee era on the *Memories by the Score* re-release (#18, the last band); Holt translates this Paul Simon chantey into reggae. With an inquisitive alto sax, it feels Kingston and Central Park at the same time. Holt's exuberance is so happy that you've got to be too, no matter what else, at least as long as the song lasts. His voice honors Simon, who is himself honoring the innocence and hope of the golden age of rock: *"Baby, I love you;/Whoa-o, oh-hoh, baby, I really really do./Lord lord, I do."*

3. "Let's Linger A While"—#26 on the first disc of *The Tide is High Anthology (1962–1979),"* this "Doo Wop lickover"* is one of the all-time silliest, most delightful songs. Holt could do pure goofiness, and here it is in spades, a throwaway love song, half-assed nursery rhyme, hash-clouded singsong, *"Please don't go 'cause I love you so;/please don't go 'cause you'll spoil the show."* Three-year-olds could style to these rhythms and rhymes.

---

*Dennis Lyons, Liner Notes, *John Holt: Memories by the Score: Classic Singles 1968–74*, London: Trojan Recordings, Ltd., 2000.

4. "Sometimes"—#25 (same disc, same anthology), this tune has a compelling madcap phrasing and beat. Holt opens like a coyote baying and later sounds like a kid laughing at his own tantrum and a troubadour serenading at his paramour's window. His lyricizing is masterful; like Sinatra it seems as if he can bend a phrase around any feeling with instinctive accuracy and unabashed sincerity. He doesn't even have to speak English or do his wonderful patois; he could hum or grunt this song and it would still arrive: *"Right now I'm crying 'cause I love you./I'm crying 'cause you don't love me too."*

5. "The Tide is High" *"... but I'm holding on ..."*—#5 on the same disc, same CD, this is Holt's most famous composition. When covered by Blondie years after the Paragons' initial release, it became a worldwide chart-topper. "'Chris (Stein, Blondie's guitarist) fell madly in love with the song,' recalled Deborah Harry, Blondie's lead vocalist, 'as did I—the musicality of it was just beautiful—beautiful melody, beautiful treatment. The harmonies on the original are very exciting.'"* Holt's late version, revised from his Paragons' hit, yearns, braves, muses, and dares—plus that's quite a country violin between verses. The cut is a stand-alone if you never heard another Holt song in your life (which is true for most people), but it is also pure Holt in its sentiments and their expression: *"I'm not the kind of man/who gives up just like that...."*

6. "Police in Helicopter"—#1 on the CD of that title, this song opens with four straight, differently chanted *"yes, boss"* cries that would do the Jolson of "Old Man River" proud. Plus the words are a Reagan-era slap in the face to the American

---

*Hank Bordowitz, Liner Notes, *John Holt: Holt Like a Bolt*, Burning Bush Records, 2001.

empire: *"If you continue to burn up the herb,/we gonna burn down the cane field."* Hard "h" on "herb."

7. "I Will"—Holt rides this Beatles classic so effortlessly that he barely ruffles its simple purity. At the top of his game, he croons another composer's work with total respect plus an offhand ease that comes from knowing for a long time that when you show up, Jimmy Cliff—and the other dudes—sit down. Holt is playful, elegant and, at moments, appropriately passionate. No one could possibly phrase *"... love you with all my heart ..."* more brilliantly, especially *"all my heart"*: his "l's" are there for only an instant, but they go on forever and, even though they roll right into "my heart," an infinitesimal gap between words says everything you need to know about man and woman on Earth. This particular CD, *2000 Volts of Holt*, offers mostly mainstream pop tunes (no doubt contributing to the diminishment of the man's reputation in Kingston), "Alfie," "My Guiding Star," and "I'll Take a Melody" among them. [Elsewhere Holt covers "Help Me Make It Through the Night," "Rainy Night in Georgia," "Green Green Grass," and "Mr. Bojangles," but I like this Lennon-McCartney (#5 on *2000 ...*) the best of all of them.]

8. "Which Way You Going Baby"—#15 on *1000 Volts of Holt* and an idiosyncratic favorite for me, this likely Terry Jacks ("Seasons in the Sun") song moved me in the early '70s when I misheard it as "Which Way You Going Billy," a variant I actually prefer. I didn't meet it again until this CD: *"You are my 'ole, babe/my heart and my soul, babe./I have nothing to show, babe/if you should go 'way."* I assume that's "whole" and not "own," but the ambiguity is both Jamaican and universal, dialectic and sememic. As always, Holt withholds just enough vocal force while getting into the song that, when a

crescendo breaks over a shift, he has plenty of room, he is all there.

9. "Sister Big Stuff"—#3 on disc two of *The Tide is High*, this composition is perfect for Holt's lyrical play. Back to back he croons childlike outrage and incipient shrillness, then switches to smooth rock 'n' rolling and the stylized teasing of so many '50s tropes: *"You think you're higher/than every star above."* Don't miss the slight hiatus between syllables of "higher."

10. "Can't Use Me" (#9 on *Police in Helicopter*)—For sheer strangeness and persistence, this Holt-composed score—like an affirmation for someone in recovery—says little more than *"Ya na go use me no more;/that we know for sure."* You have the title, you have the song. Vintage Holt instrumentation and phrasing keep going deeper into the rant. How he gets from "go" to "use" is particularly compelling. (Holt does quite a bit of minimalist playing around, especially after his return to purer reggae art; compare "Lucy and Me, Part 1 and Part 2" on *Hey Love, Hey World: "Me and Lucy gonna have a good time tonight."* There are many ad libs and variations on the basic words and music, especially in Part 2—same rough ballpark as an Australian Aboriginal sound poem.)

11. "(I'm Just A) Country Boy"—This traditional ballad (#23 on the second disc of *The Tide is High*) allows Holt's yearnful voice full range. He takes words and a tune that are familiar and wrings something mysterious out of them. As he drops into those even more mysterious minor chords, we could be lost in "Greensleeves" or "Danny Boy" and the ancient, unredeemable wars of Ireland and Africa. In fact, he's probably sung those lullabies too, perhaps for Prince Charles: *"I never gonna kiss the ruby lips/of the prettiest girl in town."* It's how he does "kiss" and then waits for "the ruby lips" and

then waits again after them. It's how this sentiment is just as real for the Prince of Wales, even though it isn't.

12. "Stick By Me (And I'll Stick By You)"—On many Holt CDs (for instance, #2, *Red Green and Golden Hits*), this ditty is playful and catchy, as its instruments range from a deep baritone sax to a xylophone. In a reggae band you could employ anything, even a piece of metal picked up on the way to the studio. *"Friends may try to hurt us,/scandalize our name;/ ... you've got a place in my heart"* is inimitable, especially Holt's trilled "in."

13. "Last Train from the Ghetto"—For #3 on *Police in Helicopter*, Holt could have written more radical lyrics; in fact, I prefer many Max Romeo numbers with the same approximate message—say, "Uptown Babies Don't Cry" and "Stealing in the Name of Jah." But the composition moves like a train, and *"Can't you see through your eyes, my people ..."* is a great start.

14. "A Love I Can Feel"—This Smokey Robinson cover appears on several CDs (for one, #6, *Holt Like a Bolt*). All through it, Holt plays off Motown and on the words themselves, with more than a touch of "Up Park Camp." If the Miracles did reggae, it would sound like this, from the bushy-tailed *"I want a love I can feel;/that's the only kind of love I think is real"* to the teasingly assertive *"...'cause, baby, actions speak louder than words."* Pause between "love" and "I," again between "actions" and "speak." Elsewhere Holt celebrates Sam Cooke with "Cupid" *(Fistful of Holt)* as well as his own "Sugar & Spice" *(Police in Helicopter)*.

15. "Pledging My Love," #7, disc two, *The Tide ...* —This is a Motown anthem from the nation of love: *"Making you 'appy is my desire, dear./Keeping you is my goal./I'll forever*

love you/for the rest of my days." In dialogue with an alto sax, it's about as earnest as Holt gets.

16. "Looking Back"—Next song, same disc, here is the perfect blend of reggae and Brooke *("What goes up/must come down ...")* Benton: *"Once my cup was overflowing,/But I get nothing in return...."* Extra syllable in "cup," withheld second syllable in "nothing," second syllable of "return" sustained, plus a church-organ keyboard on the bubble.

17. "Don't Break Your Promise"(#12, same disc)—Holt might have written this one in his sleep; he was probably singing its melody when he was thirteen: *"Now you once said you'll never love another,/and there'll never be another one like me."*

18. "Morning of My Life"—This conventional ballad (#2 on *1000 Volts of Holt*) could be played on satellite radio in elevators or dentists' offices. Yet Holt gives a scope and grandeur that you wouldn't think were in it. The mood is all over the place, and the tinge of a reggae beat never wavers: *"... watching rainbows play on sunlight/ ... in the world that no one understands/ ... it's the morning of my life."*

19. "Part of Life"—Holt wrote this upbeat quasi-sequel to "The Tide is High" and "Why Do You Hurt Me So" (#17 on *Memories by the Score*) in a slightly different voice from anything else (it might also be the funky remastering). The song has Holt's trademark hurt insistence—petulant and oversincere: *"I'll show you love/and I'll show you life./ ... I'll give you back your pride/I'll give it back to you."*

20. "Stagger Lee" (#9, disc two, *The Tide is High*)—Holt translates this legendary description of the origin of superfly/rap culture from nineteenth-century St. Loo to islandtown Kingston. A horn-heavy band dubs pimp-elegantly, while Holt's ironical narration of Stagger's murder of Billy foreshadows the

Beatles of "Rocky Raccoon."

21. "Fat She Fat"—Holt co-wrote this reggae standard (#6 on *Police in Helicopter*) with "Jungo" Lawes. The sentiments are vain and ugly, but I think that's the genre, to try to sound like an asshole—and it sure does stick in the brain: *"Not because you're fat, you're fat, you're fat...."*

22. "Happy Go Lucky" (#2 on disc one of *The Tide is High*) —This is Holt and the Paragons, early and raw, rich in texture. The band is having a pretty good time: *"Imagine how many hearts you stole ... /Everyone in town knows about you."*

23. "Sea Cruise"—Holt steps right into this old Frankie Ford classic *(" ... oo-ee, baby...!")* like a true rocker (#3 on *Memories by the Score*). The rush of water, boat horn, and opening bars from the original are dubbed, but reggae *("Be my guest/you've got nothing to lose ...")* steals the rest.

24. "(Land of) Ecstasy"—Part of the charm of this one (#7, disc two, *The Tide...*) is Holt's leaving the first "c" out of the key word: *"Ah, take me by the hand/and lead me to the land of estasy."* A kind of choppy Platters.

25. "Lost Love"—In this bubble requiem (#7 on *Holt Like a Bolt*) our guy courts with the smoothness of a Miracles or Temptations crooner, intermittently exchanging tender nonsense syllables *(tra la la la la la la, ti yay yay yay yay yay, oh oh oh oh-oh)* with a feisty trombone: *"If I were only granted/the right to start anew,/my love would be lost in you."*

26. "Up Park Camp"—This is #18 on disc two, *The Tide is High*. Not sure what the words mean, and much of the time I'm not even sure what they say, but the mood of "youth at risk," as they now call it, is all over this urban rocker: *"And me gone a o park camp;/me never wanna go a o park camp."*

27. "On The Beach"—The Paragons *"have some fun on the*

beach/where there is a party." This is a sweet Everly Brothers auguring of the Beach Boys (#3, disc one, *Tide...*, co-written by the entire group). Guys, is that *"one more box of aps"* that the bartender *"won't surrender?"*

28. "Winter World of Love"—#13 on *Holt Like a Bolt* opens like Lee Perry about to dub "Mule Train" with a skank guitar and background organ, but then all of a sudden we've got Nat King Cole roasting chestnuts on a Christmas album, and when did Holt have such a snowy *("... for love is warmer than December ...")* romance?

29. "My Number One"—Holt wrote this quirky Bunnie-Lee-era ode (*Can't Keep Us Apart*, #3 on both discs one and two) which has a feeling of Jimmy Rodgers ("Kisses Sweeter Than Wine") meets *The Twilight Zone*. Check out the polished version and then the loop. Also compare the Paragons' bouncy "Number One for Me," #4 on *Memories by the Score*.

30. "A Tree in the Meadow"—Sarah Brightman could have also covered this '40s Margaret Whiting "schmaltzer," but it shows just how sweet Holt can sound—who needs Bobby Vinton? #16 on *Memories by the Score*: *"I will always remember/the love in your eyes/the day you carved upon the tree...."* Gap between the syllables of "always."

As for the last word, JHolt: "May God bless all those who made me smile."*

The Milky Way is where stars, all of them thousands of times larger than any world, flow together in the semblance of a great

---

*John Holt, *The Tide is High*, album cover, 2002.

river cascading with the luminosity of a billion suns, so far-away and concise it could fit on a skunk's haunch.

My brother Jon hung out daily at the Driftwood, shambling a mile or so along the Post Road to share breakfast camaraderie into late morning with the old-timers. These are denizens of the real Westport, most of them raised there—indigenous, blue-collar Connecticut under the radar of upscale mansions and SUV fleets—Jon's type of guys. Following Ray's instructions, Lindy and I find the diner, as promised, just beyond the flagpole in Southport Center. As Lindy slides into a vacancy, the men exclaim variously, delightedly, "You have picked Jon's favorite seat!"

"at noon/the town/still slumbers/in grey mist//redwing black-birds/and crows/out on the marshes/call to a hesitant sun...."*

After they finish their questioning of us, they converse as old-timers do, about the past, recalling persons they have in common; they discuss their ailing sense organs and limbs, the weather, minutiae of the week and weeks past. They are all eccentric in some way or other—six or eight of them arriving and departing at different times, one in a wheelchair pushed by his buddy, one a woman (from the Pequot Library)—and all are sweet, even those that bark, doing so in a way that establishes their trademarks: querulous snipes and teases, ritual combat about the smallest details, such as who can still see across the room and who was present when Jon did this or that.

---

*The lines of poems are selected from the posthumous notebooks of Jonathan Towers.

"I've come out/onto the windswept marshes/to stroll across/ the winter-dead/short-stemmed/cowlick sportina grass/and walk among the tall/dried plumed/phragmites reeds/in antici- pation of your coming/to stand in prayer/beside the ancient tidal creek/flowing downstream to the Sound...."

Of course, Jon's exploits are the topic of the day—plus, the aftermath of his life, self-terminated a month earlier, May 4. They pass around the "cause of death" from the crematory in Hartford (self-incised wounds), commenting on it undramati- cally, with the cold, sage wit and imperturbable curiosity of the old. Then Ted Brown tells how his daughter's high-school English teacher read one of Jon's poems the other day, and a boy in the class shouted in disgust: "I know that guy; he's an asshole!" For the rest of the hour they debated this point, the consensus being that the poems told who he was, not the uncut hair, beard, and torn clothes.

"scattered boats/sit idle/anchored at their buoys ... //framed by the jaws/of the rivermouth/the Sound extends/into the dis- tance/bathed in hazy light/far as the low outline/of scarcely vis- ible Long Island//a moment of undisturbed respite/in which to feel/the infinitesimal passage/of time/and behold the perma- nent rocks/the twisting shape/of the last prolongation/of land/ and the final stages/of the river/running to the sea/like the exit- way/into infinity."

The next day, these Driftwood patrons join us and other natives and acquaintances of Jon, teenagers welcome, on a late rainy afternoon, June 6, 2005, for a memorial at Burying Hill Beach. Holding umbrellas, we read aloud from the *Tibetan Book of the Dead,* passing the text in a circle from person to person: *"You see your home and your friends/But they cannot see you/So you must be dead. /You ask: where is my body? ...*

/You feel squeezed between rocks/And tossed by the winds./I want my body...." *

The body is now ashes. Ray retrieves the box of them sent by the crematory from the back seat of his car and, heading out toward the sea, tosses handfuls in the air: they form short dust devils, smoke dissolving into void. I watch him, mesmerized and disengaged. Then he comes back and hands the container to me. I don't know why I am surprised by that. What else would he do?

I am stunned at the quantity of them, compared to the remains of Lindy's father in the '70s or those of my friend just a few years ago. *There is so much!*

"I watched two boys/playing together/on the beach/at the water's edge/finding shells/and other treasures/skipping stones/ swimming/and talking/in funny voices.... /I looked on/with that mixture/of envy and superiority/an adult feels for youth ... /wishing/I could know/once more/what it's like/to enjoy life/and the world/at ten years old...."

Squeamish at first to hold this residue of my brother's life, flesh and skeleton, memories and pain, I bear the tall box across the beach to the pebbled edge of the surf where I finally stick my hand in. Feeling the immense sticky dryness, I cup some gingerly in my palm and set its gray dust in the Sound. Ted and his two daughters crouch alongside me, so I offer the girls a turn. They participate with solemn but unfastidious enthusiasm.

From our offerings, a milky cloud forms and percolates out to sea. This is not his body, just ashes like those I scoop from

---

*The Tibetan Book of the Dead for Reading Aloud, adapted by Jean-Claude van Itallie (Berkeley, California: North Atlantic Books, 1998), p. 45.

the fireplace; though they carry a heavy symbolism. The ritual gives rise to mournful nostalgia. These are the last molecules held by the unique field of Jonathan Towers.

"puffy clouds/after the rain/float off/thru soft-blue/noonday skies/the hum of commerce/diminishes/to its mid-day pace...."

I remember us as very young children playing in the surf on the other side of Long Island Sound, our whole unknown lives before us, the mystery of our bodies so new and barely understood, engulfing and profound as enchantments of skeeball and whiffleball; sand-castle, dandelion-clover summers; sledding hills and snowmen; Almond Joys and Life Savers; that thrill, as pink Spaldeens bounce and land magically in containers for scores; 78s of Burl Ives and Tex Ritter, Casey Jones at the throttle; and complicated gameboards with their iconic tokens—the ultimate blue, red, yellow, and green of Chutes and Ladders and Sorry!.

Once I cradled his baby frame in these waters, not much larger myself, as we waited for the tide to crash over us and make us giddy; now I lovingly introduce his ashes back into the boundless and ultimately kind anonymity of nature.

"early patches of snow/remain/after the heavy rains/and though it freezes/overnight/strong red sun/in the morning/and the chirping of birds/bring the promise of warmth//the undaunted crow caws loudly/in the woods/blending his cacophony/with the brave/warble & cheep/of two sweet-sounding birds...."

There is still plenty left in the box and, after half a dozen or so people take portions in plastic bags for their gardens and favorite haunts in town, Ted and his daughters lead me along the river to the creek alongside which Jon wrote of *"things/that once upon a time/happened forever,"* things like these, and I set the remainder of his ashes in a clump where water meets shore.

Ever since, I bear a gentle but persistent, irreparable grief. My mind goes around the same cycle again and again. I feel boundless love and empathy for him and a terrible remorse that I didn't have the capacity to rescue him when he asked me to: "Come and get me, Rich." Then I remember how hard it was, how he turned on me each time I reached out to him, enraged that I couldn't make him better, jealous that I had a wife, a kid, and a calling. I realize why it happened the way it did, why it always happens that way in my family, forever.

It is no solace. He was so sweet and innocent, so gentle, even the times he was violent and a danger to me. And it is sad, sad, sad, sad, but it must be left behind. Time keeps pulling me by an irrevocable string into the now, away from here into a world that is still unmarked.

"All talk/drifts toward the sea/soft white flesh/of the brine/ tasty on our palates/fresh bread & butter/and ale/to wash it down/while our eyes/feast through the looking glass/on the calm/inlet waters/of the Sound."

As we leave the beach, I am already philosophizing, wondering aloud to Lindy how many children of the present pampered wealthy in this too opulent town, though probably not nascently poets like Jon, are fated like him to wander the streets of some other apocalyptic city?

Are we outside the Earth, like carbuncles growing on a plant, an alien predatory species invading an ecosphere once in equilibrium, thus are compelled now, as biopolitical agents, as ombudsmen (and women), as super-techies and advanced engineers, to accept responsibility for mega-damage and oversee either the planet's rescue or its final demise and death, whether we have any chance of salvaging it or not; or are we inside the Earth, part of a vast unconscious regulation mechanism, of Gaia's operating rheostat, as we stagger into consciousness, inciting growth and change, evolving despite our acts, despite our culpability in global warming, toxic pollution, and species extinction? Are we wens, epidemic urbanizing fungi, or the tree of life, the planet itself?

If the latter, if nothing we intend or plan makes any difference (because it is being planned at another level by a higher intelligence), is this any excuse for our carelessness, excess, and cavalier cruelty toward other life, land, and the sacred, which just about all politicians on the both the left and right, but mostly the right, tacitly sanction to serve short-term and selfish goals and satisfy the greed of those who are already sated? Is it that we can't help ourselves, that Earth is using us to create and destroy, to sustain and transform itself in ways so mysterious that they appear to be their own irreconcilable opposites? Are we then the real Earth, the representative of global intelligence, with all our baggage, blundering, floundering, graft, and so can neither help nor improve ourselves or the biosphere, no matter what choices we make, what measured steps we take to restore nature; or do we hold the Earth's fate in the balance of our collective political and personal actions and thus must make wise choices somehow, despite our primitivity and limitation, despite our road rage and emotional unreliability, despite the vastness of paradox, despite the universality and inevitability of conflict?

Are we on trial at large—ministers, moguls, managers, and good old boys and girls—with the all-seeing and imbedded eye of the universe our judge and jury? Or is this all heading, as scripted preatomically, archetypally, toward some Gaian jamboree?*

We are not about issues, opinions, or policies. Nothing is going to get resolved by armies, by fictionalized and massaged economies, by exhaustion or conservation of natural resources, by expansion of GNP or martyrdom of species, by windmills and electric cars, by nuclear and bioterrorist proliferation or nonproliferation. There are no viable platforms or ideas, efficiencies or sacrifices. There is, precisely, no way out. We are consciousness, and consciousness is single.

It is the conjunction of our consciousnesses that has to change, and at a deeper level than anyone could possibly imagine. At that level, there is presently no difference between George W. Bush and the Dalai Lama, or between Bill Gates and a beggar in Calcutta, between any one of us—man, woman, or child—and any other. There is probably no difference between a waterbug and an engineer. The whole kit and caboodle has to change, the protoplasmic, clairvoyant blob sunning on this great rock in space. There is no leverage otherwise, anywhere.

It will not be interdimensional flux or a Mayan calendar stop. It will not be hydrogen technology, microwaves from the Moon, or military coup. It will not be developing psychic or telekinetic powers. It will be only everything.

---

*Of course, the answer to this question must be that both are true.

The world is in total chaos because what is at our core is of a nature that none of us know about. Nothing matches itself; nothing is what it seems. Nothing allows itself to manifest. All the policies, congresses, equations, and jihads are pale approximations of a force so radical and unique, so gentle and profound, so apocalyptic and divine, so painful and excruciatingly simple, that now pure violence, cynical cruelty, ragged contradiction, inextricable bullshit and doubletalk, utter despair— our cheapness and casual exploitation of life—alone can birth it. We are emerging from the rage of the primeval sea, but it is going to be hell from here, to haul this fish out of the current and onto shore.

Those in the West don't get it about the Koran and jihad. Islam is submission. Submission to the universe. Submission to higher intelligence. Prayer. Humility before creation.

Macho swagger embarrasses us before other species, before —equally—the terrible and the compassionate, before the stars, before the abyss, before knowledge. Sanctimonious politicians, mendacious lobbyists, slumlords, me-first developers have no idea of how vast and all-seeing is the divinity they think to scam or deceive, in front of whom they parade their phony wares and cloak their apostasies.

This diagnosis doesn't excuse suicide bombings and other cataclysmic acts undertaken in the name of Islam, but it does bespeak greater millennial, even cosmic forces driving them, well beyond anyone's explicit control.

Allah
al-Rahmān—The Most Compassionate; The Beneficent
al-Rahīm—The Most Merciful
al-Mālik—The King
al-Quddūs—The Most Holy
al-Salām—The All-Peaceful; The Source of Peace
al-Mu'min—The Guardian of Faith; The Inspirer of Faith
al-Muhaymin—The Protector
al-Azīz—The Mighty
al-Jabbār—The Compeller
al-Mutakabbir—The Majestic

We asked for a global Woodstock and got al-Qaeda. We asked
for organic farms, contra dancing, and "the mind's true liberation," and got ritual beheadings. What debuted under the
sign of Aquarius—premature and callow—cast its next mark
at 9/11. For Aquarius *is* jihad. At the depth at which things
count—the bobbing of Earth through Galaxy—hippie *is* suicide bomber; nakedness, shroudedness; abstinence, indulgence;
epiphany, apocalypse; free love *is* free death. It's that far from
Kansas or Persia to Rigel and Aldebaran, locales so fundamental and different from here that their basis, projected across
the universe and onto Earth, wrings contraries through each
other such that gestalts and the shadows of gestalts merge and
reemerge, their shapes eclipsed and exchanged, as spirals shift
positions across that span absolutely more than once in meta-
gravitational calipers upon the Earth, because no one speaks
what they speak there, and no one here holds the map or attends
the parade.

In a universe this large, upon its measureless abyss, everyone denies Christ three times (at least) before the dawn that never comes. We cannot escape who we were, who we are yet to be—apostasies, fugues, and betrayals more us than ourselves. In fact, they are our ticket home. For eventually the canopy will come to town for real.

The planet and creature mind are going to be bared, not as ideology but a savage landscape scored by glaciers, an unrelenting, untagged ritual beyond rational or even sentient control. This is a geography we don't see, stranger than any cinematic rendition of the planet of an alien sun—the inside of history, the sign that cannot be deciphered, the station without a train.

If I get this spider out of the bathtub, where it keeps crawling up and sliding back down the wall made slippery by man, how many more of you are there? How many more among the worlds of night? Even if I don't get you out in time, who among spiders sorting threads in preconscious reverie will count the act, will winnow one being from another when all must enter the hieroglyph, alive and no longer alive—must pass through the blind eye of creation.

Proposal for a Short Documentary entitled "2001 Tries to Explain Itself to 1954"

To the soundtrack of Kitty Kallen singing "Little Things Mean a Lot" (though it could be almost any American pop song of that era) a Palestinian suicide bomber, about seventeen years old, strips down, dresses in pants and shirt of his final costume, records his own eulogy, prepares and blesses himself, adjusts the belt of explosives, straps it tenderly around his waist, walks out the door.* The empty room fills with a harsh, sourceless light.

"Give me your heart to rely on...."

Creatures who perish at death never existed at all. Either we are elaborate multigangliate, bi-appendaged jellyfish pulsing with phantasma, delusional matrices of an imaginary world, or the thoughts that define this existence are rooted in something utter and real, more profound and intrinsic than our flesh, denser and more singular even than the thimble of matter from which this universe ignited and disseminated itself.

We cannot feel or examine the root, but it is the focal point, the base luminosity, the original wakefulness from which we cull apperception of our existence. Without its deathless solidity to grasp at, its mantle to brace against, its inmost yogic puff to fill the night sky bottomlessly with worlds, its core pulsing unlocatably at *our* core, we would flutter vacantly and nonexistently through an even more imaginary sea.

---

*Mohammed Atta, mastermind and one of the pilots of the 9/11 flights into the World Trade Center, was said to have been an unusually sensitive child who could not bear the plights of stray cats and dogs or, for that matter, even street mice, who also refused to swat house flies. Perhaps by incinerating, crushing, and suffocating 3,000 people, he found, by enantiodromia, the antidote.

Rodney Collin, November 2, 1955: "At every death the doors between the worlds open for a while, and those that are sensitive may catch some indication through them."

Rodney Collin, August 6, 1952: "All of understanding of death is limited. In interpretations of death, one possibility does not exclude another."

Rodney Collin, April 19, 1951: "All that lives must die, all that dies must be born again."

Rodney Collin, July 11, 1950: "Death is only interesting in relation to the search for what cannot die."*

Since the planet is predominantly water—cosmic water, primordial star water—water is always somewhere.† We find it in oceans mostly, vast beyond measure, deep beyond imagination, populated by creatures of different orders than ourselves. We find it also in whole glacial continents and iceberg mountains; lakes, great and small; ponds with frogs chirking, turtles adjusting their positions on logs in the sun; rivers moving under shifting crystalline intelligence; and various pools, tarns, oxbows, puddles, wallows, rundles, and gentle rains, thunderbursts, hailstones, snowflakes, and fogs, droplet by droplet on our thirsty skin.

Water accumulates in our own neuralized bodies, inseparable from fire, organized liquidly into mind and language, approaching its own altar, made of its own prayer.

---

*Rodney Collin, *The Theory of Conscious Harmony* (London: Robinson & Watkins Books, Ltd., 1959), pp. 177–78.

†Rain and snow fall throughout the Solar System and on planets elsewhere in this and other galaxies, but many hydrocarbons are involved.

The tidepools among the rocks of Seawall are little universes—crabs, mussels, squiggling worms crossing their concourse (along with micro-seaweed clusters and subvisible animalcules). Each cosmos as such is perfect. It forms where water or wind and erosionality of rock cut a deepening depression, a cavity large enough and situated fortuitously enough in the galactic tide to hold procreant fluid. Some universes are as tiny as a puddle; others fill great crevices in the rocky wall. Elsewhere they range from the size of ponds to entire oceans, all fed by cycles of aroused comet-melt that evaporate and pour down over mineralized bodies of star-heated worlds.

Each tidepool is a Japanese tea garden, seemingly amorphous, asymmetrical, haphazard in its splashes and draining, but actually in immaculate balance, measuring exactly how much water it retains and siphoning the rest back into the tide at precisely the moment it is feeding it. Seaweeds raised and lowered by the pulse, like resistanceless lungs, receive waves and splashings and release their silts and spume, sloshing back over the edge of stone in continual drool.

What appears aimless is actually more impeccable, in deeper equilibrium than the fountains of Paris or recycling waters in a Mexico City hotel lobby, more accurate than a subatomic clock, because such tidepools use the whole universe of universes to measure their flow, form on their own, and arrive at their balancing and chemistry over hundreds of thousands of years to last additional hundreds of thousands, until they don't. Their perfection is the hardest to achieve because it can't be indexed or calibrated into existence; in fact, it can't not happen, though it takes eternity and infinite space to craft each one on any planet.

A stability that can't be measured or planned, can't be imitated, which has infinite vectors by which to diverge or become something else but doesn't because it is as it is, which literally can be anything but has computed into one thing, one simplicity and harmony, is the ultimate beauty beyond art.

Tidepools infuse our consciousness with how vast and interactive the creation is, how creatures of all sizes and shapes, with shells or without, fragile and flawless, occur where they do, without choice, without recourse, populating their zones with exactly the actions, the only rhythms and exusions, latitude will allow.

The little green snake, articulating along the driveway into the grass, has an innocent face. It is slithering across the universe in which it has found itself.

Where have we lost our innocence?

When asked how much blindness diminished his life, Ray Charles replied, "Maybe one percent"—he couldn't see the "misty moon" or grasp the delicate individuality of "chartreuse."

Ninety-nine percent of reality has nothing to do with vision, in fact nothing to do with anything, any of the thoughts running through our minds. Ninety-nine percent of the reason we want to live rather than die has nothing to do with what we tell ourselves makes us happy. Ninety-nine percent is simply here, with no perspective from which to view it, no surfaces by which to identify it, no language to reveal it to itself.

Larry King Interviews the Collective Unconscious, Which Consents to Try to Be Conscious in Order to Appear on His Show

Q. What has to happen for the human race to survive?

A. First of all and almost immediately, we need to get the global economy off oil, off nuclear energy too. Those addictions are driving other addictions and destroying domestic and community life. We need to run the world by something less toxic, more local, more long-term sustainable, probably the old 1970s combination of wind, sun, and tide, electric vehicles at the hub. That has to be accomplished in fifty years max.

Until the external flame is difficult to light, we will not find our inner source; we will burn off all our residual fuel, manufacturing squalid stuff no one needs or wants, devitalizing the planet, as we travel to irrelevant destinations and meetings that distract us from our purpose in creation. Size, wealth, and abundance for their own sake are—guess what?—meaningless.

Second, globalism has to rise to a level of consciousness and mutual aid, not oppression of one population or zone by another or in the service of cheap labor and transnational stratification. Free trade should be like the old Pacific kula ring, a goods-exchange partnership of independent islands, each with responsibility for the welfare of its own people. I don't know if we have even fifty years for this, but I'll be lavish today and give us a short sixty-five.

Third, we have to shut down the factory farms, end industrial slaughter of animals and, at the same time, return to bioregional agricultures. Plus, recycle everything—everything! Our cans, bottles, boxes, motherboards, and printed pages must be substantially post-consumer stock. That's about seventy-five years tops. By two hundred years we must be almost totally vegetarian; we can probably still eat worms and insects. The ecological cost of indulgent carnivorousness is obvious; the karmic debt, however, is deeper. It is making us crazy and turning the Earth beast against us. The souls of the animals we have killed brutally, unjustly, and antipathetically—and they now number in the trillions—are exacting revenge for their bodies we have pilfered from them. They are no longer chickens, tunas, and cows; they have allied with the planet itself and are hurricanes, wars, earthquakes, tidal waves, anxieties, malignancies, perversions, and deadnesses inside.

Next, the hyperactive reptile monkey mind has to slow down enough to recognize itself and the world it is in—shed its narcissism, clannishness, xenophobia, and assorted other indulgences.

Curtail tribal religion. The only viable spiritual practice, long-term, is a combination of Vipassana, Dzogchen, Qabbala, Sufism, and responsible initiation of the youth into the adult life—mindful attention, sincere repentance, contemplation, adoration, compassion, mentoring, prayerful supplication, and service to sentient beings and the planet. We have maybe five hundred years to accomplish that in full.

Q. Wow! That's quite an undertaking there. Do you think humanity is really ready for it?

A. I'm here talking to you, aren't I? I'm Deepak Chopra; I'm the parents of Jeffrey Dahmer; I'm Mattie Stepanek, putting an ancestral heartsong in the poems of a dying child. As the saying goes, ready or not!

Q. Evangelical Christians claim we are already living in the endtime and cite passages in the New Testament to prove it.

A. The universe is more profound than biblical legends. It

doesn't render dramas of gods with human agendas or heavens for selective behavior. It is a journey through the unconscious into the nature of consciousness itself; I am, remember, unconscious.

There is a hopeless, perverse impulse now to deny me, to pretend to know everything, to attempt to change the world by will alone, to invent gods and their speech.

Everything real, anything real comes from me and, since you cannot know me, you cannot know God. It is no wonder that those who claim most vociferously to be doing God's work resort so quickly to violence. Consciousness *is* violence.

Q. That's going to upset a lot of religious people in this country. Don't you believe the revelations of born-again Christians? One of them said recently on *60 Minutes* that the chance we are not living in the biblical endtime is less than a tenth of one percent.

A. Epiphany is not revelation. I believe those people feel the initial sparks of divine ecstasy, of their own inner transformation, and then they get scared and desperate and, instead of going deeper, thrust that unexplored feeling on a guy image of Jesus or Allah or Jehovah, and then subvert it to a secular and political program that has nothing to do with God. The intimation that one is born again is only the beginning of a far more radical and nonegoistic perception, that our beings and their shadows are driving the universe like ripples through a stream. No true path can be less subtle than the universe itself. The Dalai Lama told an interviewer, when asked by *Time* Magazine about the fates of Tibet and China after his death, "Reality will answer."

Q. I got you off track. Are there more changes that have to happen for us to survive?

A. Before a thousand years, science/technology as we know it will no longer be practiced or even remembered. It will be replaced by telekinesis and molecular alchemy. But only if we survive. Within the next five thousand years the living and the dead—the great majority as they are sometimes called—must find each other.

Q. Hard to believe!

A. This is the barest beginning. I can't make language for what will happen after that. It is too fundamental for present understanding. Let me put it this way: The emptiness that supports this entire mirage must become the actual ground of our daily life.

People don't realize what is happening. In a time of unprecedented biodestruction, weaponization, corporate exploitation, materialism, Rapturemania, and spiritual cynicism—anomalies abound: telepathy, telekinesis, precognition, remote viewing, psychic manifestation, faith-healing, Indigo children, crop circles of increasingly complex and subtle design, biological transmutation, clues to free energy, transductions of infinitesimal doses of matter, activation of cells by light palpation and projection of intention, glimmerings of hyperspace. None of these can be scientifically located, consistently repeated, or institutionalized—the UFO never lands on the White House lawn—but they happen ever more frequently and provide clues to a future polity, not only alternative modes of agriculture, medicine, and transfer of goods and information, but radically different ways of organizing economies, societies, and meaning itself.

The powers-that-be won't let go of the reigning paradigm; their vested interests run far deeper than they realize. Not only that, but the universe won't let us move to the next stage before we understand this one and the consequences of our actions. The shadow is there to conduct us to light. So the universe and the authorities are two faces of the same coin.

All manner of entities are trying to contact us—extraterrestrials, spirits, angels, ancestors, ascended avatars, Christ and Babaji in multiple forms and guises—but we are seriously not listening. From our vantage and at this distance and tilt off the main action, all sources and communiqués are more or less the same, which guarantees duality—embattled liturgies and faiths, alienation and confusion.

There isn't a real distinction between a Pleiadian and a fairy, a totem spirit and a manifestation of the Virgin Mary, a crop circle and a technological breakthrough transmitted in a dream, the Seth Material and the Course in Miracles. Information is trying to get through any way it can, in any configuration that conditions will allow, but there is a lot of static, resistance, and disinformation, so the message is effectively nullified or negated.

The dead are talking to us, mending the breach between their realm and ours, showing us where our life status and locale truly sit. Through various channel mediums, electronic and cybernetic devices, ouija boards, apparitions, and synchronicities, they are doing the best they can to be explicit. But in a circumstance wherein explicitness is more opaque and misleading than innuendo, in a world to which Christ can no longer return, where abuse and subjugation of messages are epidemic, the implicit is the most explicit mode of transmission available, even though it is intermittent and cryptic.

What do you think the esoteric meaning of 9/11 and suicide attacks is? We see political violence, propaganda, and vengeance. There is another interpretation. The living are trying to speak to the dead. The dead are reaching out to the living. At this point it is baby talk, growling, flailing and lurching, sputtering, detonating nitrates and gasolines with extravagant petulance, bumbling against the ancient barrier, a threshold that is barely perceptible. Ontologically it is equivalent to crude stone tools: mortar and pestle, bludgeon and blade. But we are inching our way there. Today's self-immolating terrorist is tomorrow's courier. He will meet himself where he least expects, 180 degrees across the cosmic wheel, coming from the opposite direction. He will do his work gently and make it back from the other side with an answer.

What else would you expect? We are illiterate clods trying to initiate a fundamental change in the relationship of energy to matter, life to death, mind to reality; there are bound to be fulminations and catastrophes. Yet that revolution is as inevitable as the one leading to this civilization was.

We barely understand the power of refined collective manifestation, but cultures work through millennia to give body to their psychic contents, to make palpable three-dimensional flying gods. All that the West has been able to summon in two thousand years of imaginal projection are relatively crude and credulous saliences of the Virgin Mary and the Savior on the Cross, but Hindu civilization has been at it five or ten thousand additional years, and its psychic libido now manifests incredible goddesses and yoginis, luminous beings with many eyes and feathers and fabulous bodies and powers that float overhead as corporeal as airplanes and clouds when conditions are right. These manifest insofar as they have the massive unitary power of the Indian subcontinent behind them.

Likewise, the dakinis and mandala deities of Tibetan Buddhism are real because a pan-Buddhist cultural representation lies behind them, sustained by group dreaming. Apache jinns and Australian totems are equally tangible to their tribes, immediate and vivid as a rock or bush.

Archetypes are powerful and not necessarily pleasant things, for they bind from within and allow no other destiny. They fill suicide bombers with such palpable realities that those who commandeered planes into buildings were imbued entirely with the immanence of a thirteenth-century caliphate.

What the Earth Is About

Do you understand this place at all? Parents in Calcutta break the arms of their children, physically deform them to make them better beggars. Gypsy girls are trained to bump into adults, create diversions, and steal their wallets, jewelry, and other accoutrements. There are whole universities for pickpockets and thieves in South America. That is the way our species has operated since the oasis and the quarry. Some people have the goods, earn them by rule of law; others prey on them. The Mongol hordes are always waiting to claim their prize, which is everything you own, and your life. Centuries of road ambushes and piracies give way to horseback plunder, train robberies, bank holdups, street muggings, car-jackings, home invasions, and street cons. Embezzlements and corporate larceny graduate to international identity theft and on-line trafficking in credit cards and social-security numbers.

I take no issue with the obvious. I note merely that this is where we still are, who we are, what our world basically is. Men walk the streets of cities between sandwich boards. The convicted are held in locked cages. Imprisoned by their elders, boys and girls assemble shoes, pants, and carpets, day in, day out, seventeen hours at a time, in makeshift factories and sweatshop basements. Is that a step better or worse than child prostitution?

From pauper urchins with twisted bodies to gangs of bandits wandering the outback and steppes of Eurasia, Central America, and Africa, preying on and slaughtering whomever, to vast international mafias with assorted goons and attorneys on call, the veneer of civilization is nothing but loot or blackmail, our various crafts, markets, and nuanced economies irrelevant mannerisms.

It is difficult even to have simple, honest friends, so much posturing and small talk abounds, self-aggrandizement and spiritual affectation. Guile and betrayal are endemic. Even the smartest intellectuals and artists carry impenetrable egoic baggage, trail hidden agendas galore. When we find someone honest and candid, with a clear heart and true words, that is special; it is the one thing that makes the world seem hopeful again. It is someone to hold onto.

What kind of structure does this place have? How do souls progress here? What do we make of our own achievements, the vanities of our existence?

None of this could really matter, the false triumphs, the victories at others' expense, and yet it goes on happening, and generations arise from toddlers and become corrupt old men and women.

There are sincere creatures scattered everywhere, threading through reality, working to rid themselves of pretensions and hypocrisies and needless acquisitions, extending genuine empathy.

It is not worth blustering or complaining about. That's how it is. In the end, no one upholds treaties, attends meetings, takes responsibility for their actions. The demise of the Earth, or at least of the present civilization, is precisely the level of its populace. That is why you can't fix it from above; you can only change its heart.

I know he composes sentimental songs with beautiful lyrics, but that does not excuse his politics or essential coldness. The Nazis were sentimental too; they cried at plays and fussed over their children like nobody's business. The overseers of the death camps and human-experimentation labs came home with candies and toys, enacted elaborate hugging and kissing rituals.

Sentimentality becomes maudlin, then turns to violence and brutality. Yet it continues to smile and show the world a genial face.

New York is packed with mammal life, street after street, every block a new tableau, a separate stage of theater and commerce: vendors with frozen dairy, pretzels, and sausages; kids kicking a soccer ball as they dart in and among parked and moving cars; a neighborhood barbecue, police invited; hobos sleeping under overcoats in doorways; girls squirting each other with a hose in a tiny courtyard garden next to statue of a dappled cow; a man displaying urban oil paintings; a chubby woman drinking from a large paper cup; suits and gals on their cells; throngs collecting at red-light corners; guys just walking, strutting, hanging out, backs against buildings; ladies exhibiting spiff and flair. There are all races, shades, sizes, colorings, outfits, hairdos, languages, decades, degrees of wealth and poverty, love and loss, wisdom and benightedness. Musicians work keyboards and drums. Dudes style perms and braids. Trains bearing passengers surge in underground tunnels, perfuming the dust. Balls and wheels roll under ordinance and free. Years of pain and disappointment mottle some visages, lodging in a hollow latency of expression, a hurt confusion of eyes; other faces light with daydreams, amusements, and glory.

This is where the ancient primate species has migrated and gathered, its watering hole, where its babies are born and young'uns play, its pubes frolic and mate. Giant stone plazas and colossal skyscraping fill vast caverns and savannas of concrete and steel, birds and a few quadrupeds feeding on crumbs and restaurant debris. Not just New York, but across the continents of the globe. There is no place else for most of these primates to go.

Barbarians, spare this city! Spare humanity! These people are vibrating with life, so hopeful, song-filled; in so much despair and barely getting by; such vulnerable, sitting ducks. They are you and me, and they need you to have compassion on them.

Please do not blast them from their urban trances and mummeries yet with your bombs. I know the cue has sounded and this is your time. I understand all this is finished, and I find it tragic beyond despair. The future, the real future is in your hands. I know you have children too, in squalid camps many of them; you have grievances that are more than just. Yet spare the city, so delicate and alive and cheerfully cacophonous in today's sun. If we turn to species war, all our dances and reveries will be lost for thousands of years.

Religious conservatives complain about the secularization of the world but, in their haste to enlist in God's legion, they fail to apprehend the actual nature of the divine. By politicizing religion, they have indeed made it into a major force of secularization.

Christianity is about the relationship between self-sacrifice and transubstantiation, not proscribed sexual orientation or zeal in behalf of latent cell clusters, not rewards for good works or vengeful banishment of the unbaptized to hell. It is the discovery of an inner Christ in the soul of every human being, not the soulless dogma of knights in armor, computer battlefields, and Christian dominion. It requires the profound transformation of heart wrought by poverty and selfless giving; it specifically precludes wealthy men and women entering heaven, however heaven is understood.

Judaism is about the splendor of creation and the discovery of the Primordial Being—the divine name—through imaginative intelligence in everyday life, through the magic of the ordinary, by seeking the lost alphabet of the Tree of Life. It is not talismanic dobbining or enforced taboos. The end to the Jewish diaspora and recovery of Israel must be internal before they can occur geopolitically. Jerusalem the Sabbath-Bride is not wed by seizing and defending Palestinian land.

Islam is about alms and pilgrimage—a transfinite spiritual

journey; it is mystical love as a mode of contemplating and revealing the unknown and multi-paradoxical God who brings us into being; it is about putting the soul into ecstatic fire and bathing oneself in the mystery of the senses and cosmic love. It is not about turning fire and breath into conquest, violence, and military jihad, or literally sacrificing yourself in acts of terrorism in order to enter paradise. It is about extending heart back into the Sun.

By disingenuously oversimplifying the path of spiritual truth, fundamentalists have lost the devout and holy in themselves and are inflicting that loss upon the world. They are devastating religious practice and replacing it with a static, grotesque caricature, modeling what their indigenous faiths would look like if usurped by Satan and imposed on mankind as blasphemies and mockeries of God.

Science, by contrast, is a delicate and fragile inquiry, a humble participation in God's mystery. When it gets taken over by its own ideologues, however, it is converted into commerce and diabolic technologies that defile and vandalize a sacred planet. Not only does fundamentalist religion stand against science, but fundamentalist science stands against science too. It scours and effaces life; it has no explanation for nature, none either for consciousness, but it subjects them to its prudish conceit and fumigation.

It should be no surprise that fraudulent academic think tanks turn against institutional science with their own forgeries like "intelligent design," as if God could be plastered on creation. ID is an attempt to bring down the secular house of science, while ignoring, as fundamentalist science does too, the truly radical, alinear basis of the universe and its divine transformation.

Science as exposition of cosmic mystery is prayer, a pure

act that is currently endangered by superstitious churches on one side and greedy industrialists and authoritarian careerists on the other. Science that honors and heals nature and blesses humankind is a sacred practice. Science that renounces God while stealing his secrets for the purpose of mercenary gain and subjugation of species is pretty much how the Devil would attempt to pervert the magic of Genesis for his own idolatry and amusement.

By one view this is a long dream—a holographic mirage. No wonder physicists probing molecules, biologists decanting organelles find naught but energy—energy and space. Remember, under a microscope anything unravels into swift-moving charges in patternings. And we consider ourselves solider than ghosts ... this realm more than a spell? There is *nothing here*. Matter is only a badge.

Fluid, porous things, we are frequencies resonating at the pitch of creation, bioluminescent blobs carrying knapsacks along the river of time—the so-called unbearable lightness of being. Alive as such—emanations, evanescences—we prowl a universe of gauze. A scientist in a dream would find nothing because its objects are substanceless. But how are dream apparitions different from transient, corruptible things torqued in nature (this arm, these fingers)? A being is a field of charges conducting templates of ancestral beads, each bevy taking on a creature's history and predispositions, shape-shifting motifs of particles under attraction, as a butterfly elapsing from a chrysalis permutes as both atoms and cells—equally dimensionless arroyos.

Everything is a special case of nothing. What an animal feels as body is electrified space—bioactive electrons and light; that's all. What it touches and desires are other translucent hives and tapestries presenting as creatures and things, poltergeists that come apart even as they come together like soot, molecule clusters giving off currents masquerading as color, shape, scent, taste, solidity. Orange is a blend of yellow and red vibrations originating in colorless electrons, their flower an oscillating chimera of particles. Words are orchestrated collisions, undulating air evaporating. That succulent aroma is odorless sparks streaming in rivulets into brain. Those locust blossoms are sunlight held in a microcosmic star.

This world is real, real and no more. That is, we can't make it less real, and we can't make it more, let alone permanent. We suffer Pinocchio's plight for an opposite reason, unable to become men and women because there are no men and women, no beasts more real.

A new sun rises, night's neons douse; reality begins again as it never was before. We can't escape ephemerality, so we live it. But don't be fooled—these are not costumes we inhabit or formal agendas fueled by decomposing foodstuffs; they are superconducting chakra webs, set alive by subtle currents streaming harmonically from every level of the cosmos.

From Otter Cliff to Sand Beach, the rocks of Newport Cove form a giant xylophone, playing the sea's melodies against the shore. All along Frenchman Bay, the theme and beat are kept by tide and waves. Musicians include: Old Soaker, Anemone Cove, The Thrumcap, Thunder Hole, and the buoy off Otter

Point, plus hundreds of other anonymous spits and inlets.

Seaweeds attached in ribbons to the rocks at Thunder Hole are pounded by surf that blasts into a tight cavity, quickly overfilling it and, with a snap or boom, ricochets back out. Combed and basted, the braids sway in crisscrossing currents, their lives budding imperceptibly.

A plant does not know that it is arising in a rambunctious attenuation of tide. It is growing into the universe, anonymous and famous to its own cilia and pores. Chloroplast-woven of emollient light, it does not apprehend anything, but it *does apprehend* reality at another level because it is alive and life is scrutiny.

If similar such vegetables had germinated in a shoreless Jovian sea, the situation would be the same. Their strands would be longer and denser, the rocks considerably bigger, tidal forces greater, cyclonic drag more vicious, the music louder, the blowhole deeper, or there would be no rock.

Habitat, or home sweet home.

It is impossible to imagine anything going on forever: history, the Earth, the Sun, the universe, ourselves, time. Because what could last forever, be accomplished in forever, have an attention span of forever? Where in this franchise, amid rabble and assault, does consciousness claim its beloved, breathe its own warrant? Where can one hide or retreat when the exhausted basis allows no egress, no suicide except into what you are, again and again?*

---

*Such is the fate of the Shades in the Underworld of Greek myth. They "suffer from an unrelieved and unrelievable sense of identity," according

Metamorphosis is the key—birth and death, sleeping and waking, being and not-being, revelation, transmigration, amnesia, duty relived of identity, new dawn. Existence is the collective manifestation of prior factors: unconscious events becoming conscious, billiard balls colliding and transferring energies, immanent forms finding expressions, formerly major stuff being totally forgotten, sinking back into unconsciousness, royalty that has been missing for millennia surfacing in different guises. We ride the Wheel of Destiny up, up, up, and then, because it is a wheel, down, down; yet we never know where we are on it or which way we are headed.

Don't even guess. Appearances deceive.

Past and future are mirages. Time is a meta-effect of eternity. Eternity does not pass in time. It needs unique, individualized cycles of light and darkness to make itself. When we are in eternity, we do not have to concern ourselves with how to spend time or what to do with endlessness because eternity is one thing, time another. We are one thing, our lives another.

Without time eternity could not specify its eternal timeless nature. Without eternity time would not happen, would not stream forth at all.

There are too many fresh ants and raccoons and ducks and ferns and seedlings in the underbrush, mullein stalks and lupine in the fields, for us to outlast them. We grow old; they are

---

to an email from poet Charles Stein: "When you are dead, you are absolutely who you are, with no chance of further possibility." That is the beauty of having a body and a psyche: one can evolve and change. Identity is not fixed.

sprightly and young, fractious, quacking, more and more of them and mosquitoes in the corridor of reality by the hour, prickly vines tangling negligently into each other's snarls.

Lichen is older and slower than time, creeping in forgotten hieroglyphs across glacial eggs. Crows are predators at festivals yet to be. Their romps across air are impossibly ardent, their ingenuousness being born and perishing each second. They are metaphors of life and death, and they are minds with wings, railing at the world with deadly mirth.

In the Tibetan view of deepest reality, after this lifetime we enter a bardo, a zone of ghosts, apparitions and, unfortunately, monsters, which block us from seeing either our own true nature or the nature of our situation. These are said to be projections that, even though more real-seeming than the matter of this world, evaporate upon recognition of their insubstantiality and essential nullity. Only then are we free of appearances to view the cosmic situation—and this is, more or less, "enlightenment," the single truly empty, unchanging state and the only condition that is not a projection.

This suggests two things: One is, that the precondition of mind rooting us in terrifying illusions and phantasm is the engine of creation. Whatever else it is, nature is an effect of this engine. And two, the world is as much a bardo, a figment domain created by projections of mind, as the bardo after death; in fact, *"bar-do"* is simply bridging—and all reality is bridgings of one sort or another. The junctions that link worlds are as real as worlds, in fact more so because states that mix departure, tarrying, and entering allow essential mind to look both ways, and at itself in between, and gander how seemingly indelible creations stand in mere relation, one to another to another, to its own passage.

Waking existence is like a dream but is actually the more truly elaborated perfection of unbroken dreaming. It is the state of projection to which a dream aspires before decomposing into nothingness, a psychic remnant without material basis, a quill inscribing upon water. Unlike a dream, nature holds its bottomless projection onto matter as unequivocally as electrons etch themselves in eternal fields in eternal space, and clouds of galactic dust and stars arise igniting at the furthest reaches of imagination and imaginal instrumentation. The lock that this projection has into its appearance is as elegant, precisioned, immaculate, inviolate, stark, and self-repeating as the chromosomal lariats that hold together the creatures that haunt this zone as ghosts.

Mind which, in a dream can only impel aliases and cartoons across an unstable landscape, in a world can set swarms of insects buzzing and throngs of birds soaring across an immense prismatic hue, entities that are also minded and recognize one another's proximal existences as absolute and unconditional. In fact, they fight each other for consumption of their bodies. On that basis of false unconditionality we, the most densely enchanted of matter-weavers, create our factories and genocides and tragedies and triumphs, great and small, never accepting that events that seem to be generating themselves outside us, that seem to be honed irreducibly of materials, are actually the result of our denial that we are able to enact and impose the fog of reality. Everything here is made and maintained by us.

To solve this is a job and a half—and, projection aside, we have our atmosphere and seas metastasizing, uncontrite terrorists on all sides, and an increasingly overpopulated, contested mortal plane. We can't get at either the source of the projection or the projector. While the biggest store of unused energy is *in* us, paradoxically the crisis has to be resolved outside in the world, not in a mind or in belief systems. This is the crux of the incarnate experience, its riddle and the solution to its riddle, its lesson and the punishment delivered until the lesson is learned. Everything gets kinked, distorted in its passage from the unconscious of the universe and of personal mind to transparency and recognizability, so nothing (Freud built a whole science on this) is what it is. Likewise, no one can come from the outside to save us—no god, crop circle, message, plan. The only choice is to go toward the roots of mind where appearances are diabolically attached and embraced by our thoughts and desires. As the Sufis say, God is as close as your jugular.

Conversely, going toward the attachment of mind cannot disperse or cure the shroud of illusoriness because *this is not a dream*; it is a far more substantiated and subtle appearance. Realization is not about rejecting and trashing projections (they are irrenounceable); it is about accepting them, unafraid, *as what they are*.

The movement toward the root of mindedness leads to compassion for other sentientized beings in place of fear, envy, defendedness, predation, contrivances of revenge. Getting on a path of nonattachment, even its first hesitant gestures, releases such bounteous and bottomless waves of compassion that our very place in the universe shifts and we are in awe of the depth and majesty of creation and accept our fate, whatever it is; compassion then becomes emptiness. Instead of wanting to cache and hoard, we want to share. Instead of trying to liberate only ourselves, we mean to set everyone and everything free. The paltry and menial are replaced in a breath by the cosmic and serene because these are one and the same at different phases. As we felt incipient anger and fear, we feel rhapsodic sorrow—almost unbearable, healing sorrow. Instead of wanting to acquire, we want to serve, not out of altruism but from our naturally manifesting, innate goodness, from objective understanding that giving is the only way out of the trap. Though it is the merest baby step of liberation, it a huge one. If everyone in this world took it simultaneously, all problems would be spontaneously solved, their symptoms universally eased. And we'd still be in a bardo. The granite cliffs would still stand. Far more profound meditations lie between us and their essential molecularity, between our mind and theirs.

This is the dilemma facing us, man and beast, after our deaths too—for the conditionality doesn't change, only the costuming and locale. We continue to be at the mercy of mind spewing thought-forms, utterly convincing, corrupting, and corruptible. Probably we can't improve the journey; we can only go on it.

There are many choices every day, but only one choice.

And anyway, these are just words, more ephemeral than reality by far. The event they purport to depict is quite different from their depiction of it. Famously language is an approximation—through the semeiotic logic of the brain and the neural pathways that underlie it—of a state it arises from and cannot encompass or comprehend. Words come to their premise or conclusion antiseptically, abandoning us in a jungle that is a billion times frothier and more dangerous, more deliciously and irrevocably felt. The best intentions go for naught, and enlightenment becomes an impossible distant goal when nausea grips the belly or arousal the loins, the bees are summoned into the mating dance, the executioner forces on the hood: "Tapout denied."

We are as unstoppable and undeterable as hyenas in our rampage on the forms that feed us our energy and assuage our continual hungers. Depression and anxiety are tireless birds of prey; desire and pleasure feed likewise.

But that does not make it hopeless, for appearances are ever sustained by a deeper reality, immune to mere assaults and changeless in its muezzin.

Embodiment is exactly that—a multidimensional knot of spirals that keeps the beast in the labyrinth, the snake in its snarl. Neither the cage nor the beguilement are real.

Let us pray, not to tribal gods who provide only attachments, false and prideful triumphalism and xenophobia, good against evil, them against us, but to the dancing spirits within as they hold nameless truth in their generosity. Let us pray to be liberated into boundless mercy and love.

The death projection is false. Someone else wrote a story and built unauthorized scenery. The universe provided masks and symbols, as it always does, for everything. No one put on the livery of the grim reaper. There are no goats, even here, even there. No smith or welder exhumed the fires or fashioned the irons of hell. No totem eagle or abyss of zero waits to annihilate you.

While crossing Western Mountain, this warning issueth to seabirds, clams, peoplekind, porcupines, chipmunks, beetles, baby raccoons, and the like—avoid the lamps of habitation, the vines of desiring, the switch-backing fog, the towering mirages of rock—the smoldering smudgepots of Earth. These may look and feel like sanctuary, like home, but—beautiful, oh so beautiful and fragrant, mother, mother—they are deathly phantoms, magnificent baits of illusional mind.

Go where you most resist going, where you are least coddled or charmed, toward the vacant expansiveness of sky, sky within.

The chant of the Dalai Lama comes. It is as thick and clean as this manifestation itself. All things are in it, birds, memories, light, sorrow, the clouds that are dissolving as sun clears the horizon. Mist or not, it is all mist. No matter that this plane is opaque and rock-encrusted, vast and radiant in its billion-fold night, populated as an anthill and the march of refugees across a starry desert, we are here because we have arisen here, along with our bodies, along with our attention.

The custom is to call it the universe but, as I begin my seventh decade in this body, I know without being told, know in my bones, that this is merely a fragile vibration at an unknown frequency, one of many. Men have male bodies; women have female ones; I have a man's shape this time, and his arms and legs—but it is a garment nonetheless. With the Dalai Lama's prayer, all worlds are rushing like the clouds across the sky toward the center of all vibrations and chants, toward a fixed point from which everything emanates and takes shape, pulsates to make atomic reality.

The feral kitten who appeared outside our window during the passage of the last remnants of the hurricane, wet and frightened and squealing, has a body, is another visitor to this frequency, hears the lama's chant in its bones but doesn't hear it, arrived only a few weeks ago but is here forever now, at least today. It is real, but it is not real. It is as real as anything gets. And we need to rescue it even though god knows how many more creatures are alone and crying in their vestments throughout the worlds of the universe; this is the only one we can take in tonight. It is grateful, dries out, purrs. It lives here now. The chant goes: *"Sugandhim Pushti Vardhanam."* Let us follow kitten and lama from world to world, immortal, unafraid.

In 1996, after our thirty-year college reunions in Massachusetts, Lindy and I made a pilgrimage back to Mount Desert Island, twenty-seven years after we first arrived there from Ann Arbor to begin my graduate-school ethnography of Maine fishermen, and twenty or so after we took leave of the island, seemingly for good, by moving to California.

Back in '69–'70, with our newborn son Robin, we rented a cottage on Indian Point Road off Green Island Landing. All that winter, while I conducted interviews and analyzed half a century of fishery statistics, through the last fluffy snows of April, we searched for culture beyond lobster wharves, clam flats, shrimp factories, worm cellars, bait sheds. Yet it was a world of fishermen, carpenters, mechanics, masons, clerks.

A rowdy Dartmouth graduate did lead us to the Northeast Harbor home of Marguerite Yourcenir, a French novelist hibernating with her companion. Later, he put us in touch with Bill Moise and Eva Reich, son-in-law and daughter of Wilhelm, the late radical therapist and orgonomist (they lived near Winter

Harbor on the mainland, just off Mount Desert)—but I was too young and reticent to grasp either the historiographic grandeur and European lesbianism of the former (and Madame Yourcenir was not yet famous enough for me to suspect more than an imperious eccentric), while Eva and Bill scared me off by their inquisition into the purity of my devotion to her father's martyrdom—their questions were belittling and aggressive and had no right answers anyway.

Mount Desert was the end of the earth then, an outpost of the mysterious downeast, the Vinlandic land-bridge the poet Ed Dorn had dubbed "North Atlantic turbine," mythic and inscrutable, like the Tlingit grange hall I visited with him outside Anchorage seven years later, the elders' sly diffidence betraying nada. At the juncture of Eastern and Atlantic time zones, October waning, the small sun behind old mountains casts penumbras of premature darkness, the shadow as internal as manifest. Lindy and I were a young couple then with our baby, on a tremulous shoal between our just-completed childhoods and our children's yet to be. Our life together was bouncy, unprobed. When I smelled spruce and pine and delicate incense of beach roses, eternity and the now came together in an epiphany of companionship and romance fulfilled.

I feel that texture so much more poignantly now, over a decade after our other child, Miranda, outgrew our home in Berkeley and, rechristening herself Miranda July, moved to Oregon to become a performance artist and film-maker, leaving us an empty nest, a lone couple again. Under fast-moving weather systems and noisy conclaves of crows and gulls, Mount Desert transmits a forgotten incarnation on a distant, beautiful planet—both our own and our children's childhoods now come and gone.

Agua sparkles in breezes over Jordan Pond and Eagle Lake. Cumuli gestate into swift-moving castles driven by high, invisible wind. Waves deposit their pagan symphonies onto Seawall, each unfinished, each lost forever. In early autumn, Echo Lake, fed by rivulets down Beech and Acadia Mountains and Saint Sauveur, chilled by a harvest moon, is not too icy for immersion, inviting five-minute swims amid red and gold litter on the shoreline current. Ducks and gulls ride the canvas, clouds torn to shreds overhead, liberating absolute blue.

Splash! The wonderment of being a water creature again, as skin and viscera adjust to immersion. Memories of all lives, recoverable and not, peal in the limpid shock of cold on skin, neurons converging on something so direct and profound it is shocking to be alive. I lie on my back and make starfish movements to propel my body, arms and legs extending and then drawn back to torso in a dance more ancient and fundamental than me.

Triturated by wind, tuned by crystal, vitalized by solar glittering, the lake is wooed by timeless minor thirds from the loons—plaintive glissandoing calls that bind the water to its human name. Echo seems potentized to at least the decimal.

Mount Desert is a magical place where people risk more than they intend and are granted unexpected indications. Mountains, pools, and caves hold vestiges of Old Earth; woods and their habitants hover at the edge of a fifth dimension. While hiking, people meet others connected to their lives in uncanny ways. Ravens, foxes, squirrels, snakes, turtles, darning needles, spiders, one by one by one in inexplicable encounters, know more than they have a right to and, though languageless for generations, find ways to communicate ineffable, immortal things. From Somes Sound to the Bubbles and Flying Mountain Loop, along the cairned trails of Acadia Park, semi-intelligent creatures cross trajectories bearing messages or gifts, even if only once or incidentally—concordances at such unlikely frequency it would seem that signals from Cadillac and Sargent alone could be scoring them, steering us lonely mortals into each others' paths.

Island, thy name is Fortuity.

By 1996, enough back-to-the-landers, New Agers, and dropouts had been attracted to this shire that, at times and in spots, it felt more like Berzerkeley than Berkeley itself did. Alternative cultures had grown up within towns like West Tremont, Somesville, Bernard, Hulls Cove. The island had its own institution of higher learning, College of the Atlantic, an old seminary on hills outside Bar Harbor, co-founded by Father Jim Gower, pastor to fishermen and environmental warrior, an activist more Christian than his church. COA granted a sole degree, in human ecology.

Notables of the mid 1990s included P. Chris Kaiser, trail guide, amateur shaman, and aficionado of the vision quest; Richard Handel, proprietor of Eden Rising (spiritual artifacts and books) and master of ceremonies, greeter, philanthropist, professor beyond portfolio; Bill Kotzwinkle, author of *Fan Man* and *E.T.: The Extraterrestrial*, the island's mage, hermit, and spiritual thug, cultivator of invisibility, who could pass through the middle of a busy thoroughfare and not be there; Ed Lueddeke, magister of the firewalk and proprietor of a unique energy system featuring an invisible electromagnetic field of complementary light; Barbara Thomas, reiki master and Great Mother; Paul Weiss, *chi gung* teacher, healer of lost souls, founder of the Whole Health Center off Crooked Road; Aubrey Barr, blues maven and closet novelist, late of New Orleans, Southwest Harbor school-bus helmsman; Steve Perrin, uncoverer of buried trails and tracker of archaic crabs, public conscience against Jackson Lab's spreading invasion of the Park under the decoy that raising onco-mice fulfills its environmental waiver; Hawk, Amerindian emissary to indigenous Australia and Peru; Masanobu Ikemiya, cosmic refugee and transcultural musician. There were others, too, including the spouses of many of the men and a variety of characters who came and went with the seasons. Even Wendell Seavey, star of my thesis, one-time boss fisherman at the Bernard wharf, an alpha male in a rough neighborhood then, had since cultivated his psychic side and, toting his magnetic eagle feather blown onto Lopaus Point, now parleyed with spirits crossing the island, taking direction from the ancestors, and saw which birds weren't really birds, having transformed himself from a party animal and redneck-in-training—the legendary "no good boyo"—to the local equivalent of ecowarrior and master of the tea ceremony, geomancer of fog and wind.

My old writings—*Book of the Cranberry Islands, The Provinces,* and *The Slag of Creation* (as well as my Ph.D. thesis, on microfilm in a few town libraries, *The Strategy and Ideology of Lobsterfishing on the Back Side of Mount Desert Island, Hancock County, Maine*)—had imparted some status on me *in absentia* as a pioneer of local cosmology. Since only Wendell still knew me in person, he extended the invitation.

I first learned of Chris Kaiser when his then-wife Karin, upon hearing that we were planning a visit in June (and knowing we were publishers), introduced herself by post that April— an unabashed request that I turn her husband's autobiographical manuscript into a bestseller. Since Chris and Karin were Wendell's buddies and I wanted to get off on the right foot, I managed tentative enthusiasm.

Unsolicited manuscripts, especially ones about spiritual journeys, have been a bane—mostly literary neophytes narrating "my life, beliefs, and encounters with God [or the supernatural]." Chris' version turned out to be as blatant as any, rivaling dreams of dying pigeons and prophecies for the twenty-first century channeled through Marilyn Monroe. Entitled *My Heart is Sacred*, it wove heedless assertions about himself and the cosmos into grandiose myths. Yet something ingenuous and gullible moved me more than a hundred or so of its predecessors, most of them better written. The truth is: we all live such myths; the dilemma is how to claim and inhabit them without overstating or oversimplifying, without slaking their delicate magic and transpersonal mystery by requirements of literal demonstration.

Two subplots of *My Heart* ... stood out. One was Chris' tale of how his two-year-old son, Wolfie, had contacted him years before his conception and arranged the love affair that led to his reentry onto this plane. Wolfie had been an Apache Indian chief/medicine man named Stalking Wolf in a previous lifetime and selected Chris as his mentor and guide this time around. The second was an account of "the chakra system of Mount Desert Island," compiled while hiking Acadia, sleeping at its power spots and attuning to their apparitions.

Chris had fashioned a geography of sacred sites beside which my *Book of the Cranberry Islands* was post-modern jargon. His text belonged in a library of indigenous geomantic works alongside Samuel Champlain's humble description of blue bubbles afloat in an untracked seventeenth-century sea: "This island is very high, and cleft into seven or eight mountains, all in a line. The summits of most of them are bare of trees, nothing but rock. I named it *l'Isle des Monts-déserts*."*

In Chris' *Heart* Champlain's sierra archipelago came alive as a Sedona-like power destination: mountain deities and gnomes, pools bearing luminous salamanders and goddesses, medicine wheels cast in moraine, tarn chakras, kettle-pond *nadis*, and ghosts of Red Paint People piloting canoes of astral birch. A split boulder in Acadia Park projected a crack in the cosmic egg.

Of course, there was no way we would publish such indulgent New Age tripe. But, as usual, fortuity intervened.

The following summer, while spending a week on Mount Desert, Lindy and I attended a lecture by Chris in the space above Eden Rising, a reclaimed attic which had recently featured a yogi, a breath-worker, and Hawk on his didjeridu. After the talk Mr. Handel laid out the plan.

The next afternoon, hauling a tiny Wolfie in backpack, Chris guided R. Handel, Lindy, and me from Cadillac summit—the winding road up which we had just driven—a mile down its South Ridge. As we clambered along ledges, sampling tart blueberries, no patch tasting the same as another, Chris pointed to neighboring mountains bowing before Cadillac and drew with his hands implausible feng shuis and megaliths that titanic ice creatures had disposed millennia earlier. Winded by the hike, we were in a respiratory enough state to receive the tender vibration of a cordate pond set a plateau of ledges beyond our path's terminus, gateway to the heart chakra of Cadillac Mountain. With its margin of reeds and cattails, it opened our hearts and,

---

*Samuel Eliot Morrison, *The Story of Mount Desert Island* (New York: Little, Brown and Company, 1960), p. 9.

at that moment, we glimpsed the foothills of the entire trans-dimensional system. Wolfie tried to pee, but a stiff breeze off the sea blew his offering back. He said, "Someone wetted me."

By the following spring *The Chakra System of Mount Desert Island: A Guidebook* was a reality. Our abridged and edited portion of Chris' opus began with local geology and history, while establishing the principles of all sacred geographies. Next it provided sets of classic walks: the solar plexus (Eagle Lake), the sacral chakra (Valley Cove), the base (Western Mountain), etc. Each major mountain also ruled its own subsystem such that, within the microcosm of Cadillac itself, the base chakra was a previously uncelebrated chunk of moraine that Chris had named Wolfie's Rock. "The size of four cars stacked two on two, [it] is located about forty yards to the left of the trail. There is a circle of stones in the ground on the far side of the rock. If you stand in the circle and face the rock, you will see a face."*

After Wendell's adult daughter Mary had developed a brain tumor a few years earlier, Chris led her to the Wolfie Rock for its intervention. He and Wendell believed that, though she had sound reasons for wanting to cross, the wise boulder kept her on this plane a bit longer to prepare for the ultimate journey. WR later visited her dreams, in one of which a bear approached, picked her up, carried her to its altar, and placed her on it "at which point she felt a soothing energy run through her body."†

I was introduced to WR when Wendell led Lindy and me a mile down the South Ridge through the Black Woods. There

---

*P. Chris Kaiser, *The Chakra System of Mount Desert Island* (Berkeley: North Atlantic Books, 1998), p. 105.
†ibid., p. 110.

it stood, a stone's throw off-trail, sphinxlike yet sincere among fellow immigrants. The hall-of-fame fisherman, now Chris' partner in clairvoyant crime, pointed out where Mary's ashes had been scattered and then where he wanted his own to go, in the crypt under the tomb, to return (he said) some nourishment to a creature who had given mortals so much.

In 2000 Lindy and I bought a cheap house on 102A in Manset and began living in it a few months (late summer/early fall) each year. Chris was one of the people we got to know well. Before dropping out, he had been a salesman in Massachusetts—two older daughters from former relationships still visited him. A towering imp in his fifties with curly hair and a pixie gaze, he was a combination hippie/triathlete. Dripping sweat, he would pump his bike ritually up Cadillac's tourist highway, a world-class marathon. Paced by imagined Indian guides, he also ran grueling miles on Acadia's roads and trails and bathed in the glacier-melt North Atlantic longer than any nonmarine mammal should. He mostly butterfly-stroked a mile or so across Echo Lake and back; then, without resting, did it again, often as not a third or fourth consecutive time. As I frequented the Ledges on late summer afternoons, shyly inching my way into the hibernal waters, the sight of Wolfie and friends splashing with plastic shovels and pails or running along the cliffs meant that Chris was out there somewhere, Big Warrior embracing ego-death, with a cargo of compulsive self-flagellation.

He earned wages hiring odd jobs, mostly carpentry, but also rock walls, painting, clearing ground, hauling debris, laying out bocce courts, assisting in general construction and repair. Self-taught in t'ai chi ch'uan from a book, he conducted a celebrated morning class in which many of the same tourists (as well as locals) enrolled summer after summer. Once *The Chakra System* debuted as a signature green tome, part of his routine included hawking copies store by store, mostly in Bar Harbor, and organizing adventures for sightseers weird or perspicacious enough to track him down.

The Park shop wouldn't carry the title; in fact, assorted Acadia bureaucrats, who thought of Federal land as an extension of the Stars and Stripes, were scandalized by Chris' patent invitations to vision questing within its boundaries, especially his championing of ritual fires and overnight vigils, an arrant provocation. Most of the millions of annual travelers to Acadia wanted secular scenery anyway; a rare hiker preferred the Dreamtime.

In addition to ley lines and mumblings of Taoists, more troubling spirits haunted Chris—specters from a paranoid universe that portended mind control and satanic takeover. A late exuberant cosmic hippie, he was not otherwise easy-going or contactful. Beneath his affable veneer and peak performancing, there was a tortured soul, lethal moods and bottomless sorrows kept at bay by a regimen of New Age vigils. Distancing friends from any real intimacy by spacey raps that careened from reckless optimism to deadpan apocalypse, he was finally left with the island and wraiths he invoked—bewitched as well by woman magic, (often) young females, one after another, who enchanted him by, perhaps romance, but more crucially an alliance with the invisible world, as Wolfie's Mom once did.

Rather than engage his own misery in common-sense or psychological terms, Chris went out daily questing, putting his body through more and more rigorous asceticisms and challenges, determined to rid himself of any outward glumness so that indigenous spirits would recognize his devotion, intercede, and grant him his elusive happiness.

To most of his acquaintances, the chakra maps were whimsies or metaphors; they never took them (or Chris) seriously at the level at which he proposed himself. His system was honorary and recreational more than essential; he was humored as a keeper of neverland, a custodian of were-clowns, a harlequin of magic spells.

Others intuited gods on Mount Desert, wavered, as Chris did, between hermetic naiveté and New Age ambition. Yet he achieved the epitome of this dilemma because he was desperate, hell-bent, and pure. This gentle, arrogant guy so literalized totems that, in the absence of social awareness, they became promiscuous merchandise. Many in the community knew them by heart and spread them along with the latest tall tales of Chris' adventures in chakra-ville.

Like many New Agers, he had privatized honest intuition, an early clue to something unknown and wonderful (Mount Desert's web of fortuities and esoteric vibrations) and made it into a flagrantly self-serving Celestine Prophecy. Without grokking it, Chris had suborned himself, sabotaging the mountains' simple integrity by his need for hyperbole and instrumentality. I mean, conversant bears, zap rays from hill gods, waves of vital force pulling him into the astral while his body slumbered alongside the chakra field of a pond, a sheared boulder marking the "crack in the cosmic egg"—pure wannabe New Age! What should have been vajra alpine tangkas became florid Candy Mountains. What should have been the soft bottomless pools of his own sorrows were bluffed, by stubborn invulnerability and counterphobic reversal, into basins of metaphorical chi. Uncontacted trauma and loneliness were rebirthed into springs of yin-yang lemonade. He was going to heal himself by shamanic action; nothing else was worthy of a Red

Paint. Perhaps more to the point, he was never going to revisit the dungeons of his soul or the American culture of death he blamed for his misery.

Richard Handel was a totally different sort of seeker. While he promoted Chris, found him clients to gaggle through Acadia, and good-humoredly inflated his oracles and adventuring, either he took nothing that seriously or took everything so seriously that it was all "chakra systems."

The Handels' house on the Ripple Pond marsh outside Somesville was the island's dojo. Emceed by Richard himself, epic ceremonies and encounters took place there every summer—fire rituals conducted by Tibetan lamas, community rebirthings, wakes for the recently ascended, potlucks with Boston chiropractors and Eastern European alchemists—even the handing over of Bill Kotzwinkle's raw (but soon to be famous) manuscript of *Walter the Farting Dog* to Lindy and me at a September 2000 dinner. This piece of inspired Handel networking led to a national bestseller and a library of "farting" books, a cornucopia that the shopkeeper in Richard foresaw from his experience selling *Fart Proudly* by Ben Franklin to tourists years earlier. When I told him, in pre-*Walter* days, that there was nothing funny in the Franklin book—and nothing even really about farting—Richard scoffed, "No one cares. They like the title. They don't even open it."

Walter became more than a zany children's story or publishing success. It generated a series of lessons for Bill and me, some sweet, some bitter, about fortune, friendship, ambition, betrayal, and forgiveness in twenty-first-century America, as Richard predicted it would a few days before his inopportune death.

It was at a gathering at the Handels' that young Rinpoche Thubten Tulku transmogrified another clever plot of Bill's, about a bear who stole a manuscript and became a celebrity writer, into a parable about a similar bear who stole a manuscript in order to become enlightened, enlighten the world, and then return to pass his teaching on to the author. (At that same event, Richard Handel interrupted Thubten's solemn succession of mantras to say that he thought he heard "Jingle Bells" in one of them. Thubten proceeded to chant *"Jingle all the way...."* and then suggested that what would be manifested soon were presents and Santa.)

Sparks flew when Indian pundit Shanji confronted the American faux-Buddhist teacher Byron Katie. After Katie held a flame to her hand to demonstrate that she was beyond pain, Shanji grabbed her fingers and twisted them in order to show that she was faking. "Did you see her grimace?" he declared. "She is trying as hard as she can to suppress that it hurts like hell. You Americans are so gullible. My people have been doing the guru scam for ten thousand years. The lady is full of shit!"

In 2004 Bill joined biovalent therapist Frank Lowen and me hiking up Beech Mountain and, as we paused on ledges overlooking Long Pond, the fjord and isles and mountains beyond, he told this story of the Dreamtime before there was a world, before there was anything at all but space, when man wanted to be made manifest. He asked God, Allah, for a mode of expression. And God loved this creature so much that he made a spot for him in the universe the only way he could, by giving his own body as rock. He became perfectly still so that humans would have something to stand on. "Can you imagine the sacrifice involved," importuned Bill, "turning himself into stone, consigning himself to such depth of immobility and silence forever, that we would have a planet? I think of that every time I walk on these rocks; I am walking on God's face."

Frank remarked later along a parallel track, "There are some people who meditate so deeply that they reach the point where they see everything in all creations. They know how to change *any*thing. But no one who gets there ever does."

Mount Desert, in the guise of a recreational area and tourist mecca, is, as Chris suspected, sacred business for those who pilgrimage on its body—land of undines and trolls, mirrors and dreams, exile and redemption. In the social register it may be a series of yachting harbors (Martha Stewart and the Rockefellers have built lordly estates among them), but those are ephemera and props and, underneath them, beneath Chris' sacred geography as well, lies the old country, no catered regattas or astral menageries there, instead the urgency of dharma, claws sharpened by the prankish trickster fox.

This is what Tolstoy called "life itself," a state from which we tend to shrink, preferring the "reflection of life," officialdom. We experience "a feeling akin to that of a man who, while calmly crossing a precipice by a bridge, should suddenly discover that the bridge is broken and there is a chasm below. That chasm was life itself, the bridge that artificial life in which Alexey Alexandrovitch lived.*

MDI gives out practices, hard yogas, whatever shape these take for each person. With sudden shifts of light and storm, the energy endemic to stone (massive and fractured) strewn

---

*Leo Tolstoy, *Ann Karenina*, translated by Constance Garnett (New York: Barnes & Noble Classics, 2003), p. 134. An island is a perfect place for reading or re-reading Tolstoy, Dickens, Hardy, Edith Wharton, Dostoevsky.

with prodigal indifference, serenaded by pagan crows, the land catches you exactly where you are stuck and corkscrews you in a direction or at an angle you would never go on your own. Many of the rusticators sense these lessons and are summoned thence unconsciously by their omen.

Lamas, sheiks, and their students apprehend the priceless clear-mind transmissions, the menace and protection of gods of the mountains, who are dispatching, at sonic speed, a collective, blind, fundamental dzogchen, which hits each initiate in a way in which he or she is ready for it, or not. The sign on the Trenton causeway should read: "No fickle initiations, ye who enter here."

The slippery rocks under Long Pond have taught the soles of my feet the painful unleashing of stale energies by jagged stone shiatsu.

Allowing my eyes to open beneath the Ledges of Echo Lake, I have glimpsed ghostly reeds and underwater moraine in the vast muffled light of all mortality.

Palpating boulders on Parkman Mountain as if diagnosing them osteopathically, I have gauged how deep and vibratory the base of creation is.

Getting lost and then finding the trail again amid cranberries and spider webs on Sargent, I have felt my entire adolescence return in a single relieved rush.

Critters thump and scuttle along roofs at night. A tire on a rusty bike bought at a yard sale explodes in my face as I run an air hose into it from Gordius' automotive kiva, a terrifyingly visionary burst of soot, stars, and tintinabulation, a suggestion of how life might end abruptly at a Tel Aviv café.

Rogue waves at Seawall tease me with their Homeric reach, and roll back into eternity.

Along the fern-camouflaged Indian trail at the base of Beech Cliff the ground suddenly gives way, and I fall up to my waist into the past.

In 1969 outlaw fisherman Buddy Lawson locked me in his cabin on Duck Cove, threatened to shave off my long hair and beard (of that era), and then offered to release me if I shared liquor from his bottle. So we exchanged tall tales, competed in boasting, bleated about women, and cursed out our favorite assholes, bonding pure male energy, something I had managed to reach my twenty-fifth milestone that night without experiencing until an alcoholic truant made it my sacred birthday gift.

Fancying myself minimally adept with a kayak in 2002—the summer after the autumn Lindy and I bought two used ones and got a quick lesson from Frank Bauer at Acadia Park Rentals, the talkative proprietor who claimed he was retired CIA and had armed Osama with Stingers during the Afghan-Soviet War—I proudly paddled through the birdbath of spattering gulls in the middle of Long Pond to the far shore beneath the rope swing and, with rolled-up pants, stepped out into what I thought was knee-deep water. It was so deceptively clear that I sank fully dressed, wallet and all, up to my chin (cell phone as well) and, simultaneously, the kayak turned over and filled with water. I had only my cupped palms and a flat rock to scoop with for over an hour.

In the summer of '05, kayaking the more inhabited side of Echo Lake with my brother-in-law, one-time Marine Jim Doyle, I seconded his spur-of-the-moment suggestion that we check out the distant shore. Three pulls on the right half of the oar glided me into the exposed middle. A wind stirred up. As I proceeded at this new angle, choppy waves suddenly spat over my prow, rolling the boat back and forth until I was nearly too seasick to paddle.

It was an existential moment. I was dead center and, to the lake, no more important than an afflicted gull or ant marooned on a stick. I was also a land mammal, and I had to get my miserable raccoon-beaver body back onto terra firma from only who and what I was, an exact inventory of stamina and bare skill I carried on board with my shirt, floatation device, and keys.

I rowed for dear life, stirred by an imagination of Davy Jones' locker, the grave of turtles and possums, viscera fallen to Echo's uncharted depths.

Ten minutes later, dizzy and weak, I stumbled onto rocks alongside a brief rotting pier and threw myself on its boards, sinking into primordial trance.

Awakening even queasier, I stumbled into the lake in an attempt to dissolve the haze medicinally. I floated on my back, then dove under muddy lymph that had cascaded and filtered downstream from all the wooded crowns encircling us, eyes open, caressed by ghostly reeds, a dreamlike miasma inside which nothing seemed solid or stable, my limbs no more mine than those underwater stones.

Finally, my nausea unabated, Jim proposed, to get ourselves home, we proceed along the shoreline, in and out of coves, circling the entire lake if need be, to stay beyond the breeze.

Once launched, however, he saw an alternative. If we picked the perfect diagonal, we could abut the chop and not have to rock in it.

It was arduous rowing into waves pitching us head-on, surges of vertigo, the shore getting closer by infinitesimal degrees [a journey I dreamed a week later, on my back on a raft in the ocean, having drifted far out into the deep, the sun of child-hood pouring from a virgin sky, a seaplane landing somewhere behind me (but I couldn't open my eyes to see it). A spirit came up from the waters, clenched at my left arm, then tumbled over me, back into the sea.]

"At Ikes Point," Jim told Lindy, "he jumped out of that boat and collapsed in the sand. I thought I should throw a blanket over him, or maybe some dirt."

It was good to go on a mission with Jim—unintended but appropriate that the mission was me.

The island provides tea parties too: wharf horseshit around woodstoves, blueberry-pancake community breakfasts, pop-overs overlooking Jordan Pond, home-made root beer and raisin buns at Little Notch, and $6 first-run art films at Reel Pizza. Yard sales, beginning Saturdays as early as 7 a.m., are more like potlatches than backyard capitalism. For a dollar or so per item, I have bought not only an abandoned husband's two for-lorn bikes, but squat oval rocks painted into comic, almost manga beetles and turtles; recycled stuffed animals (including birds, foxes, insects, and speckle-bellied bears); hand-tinted floral mandalas on wooden plates; 1950s dictionaries and ency-clopedias fat enough to fill eight inches of shelf space; odd garage and kitchen utensils such as massive files and restaurant-size colanders and pots; a giant, glossy papier-maché of a startled rabbit; and fungus-covered paintings of wharf landscapes by someone's grandfather (the fungus, wiped off with a damp rag, leaving a curiously modern surface suggesting Seurat).

Then, in the words of Paul Weiss, "on December 5, driving into a snowstorm on route 81, south of Binghamton, New

York, Richard Handel's van skidded across the median into an oncoming tractor-trailer. There is every reason to believe his soul left the body at that point, although his body was alive a few hours longer....

"It was a consciousness-raising event as everyone began to realize what they had taken for granted—for he had such a sleight-of-hand way of keeping the attention on others, but ending up under our skin and in our hearts....

"Our biggest shock, our biggest indignation perhaps, is that Richard appeared to us as the eternal host of everything; and the host is not supposed to leave before his guests. We are awkwardly abandoned to ourselves, just as he was always pointing to ourselves in the first place ... his sometimes mysterious, sometimes just beaming smile...."

Richard was the grand dragon of Mount Desert. Unassuming, even self-effacing, he bestowed generosity, grace, and the wisdom that comes from living comfortably in one's own heart. His continual koans were creatively insulting, as he teased people about their weight or love life or failed career, whatever weak spot they might be favoring. At the same time, he inflated everyone embarrassingly into a famous author or musician, volleyball master, the best body-worker in New England, to cite a few, ditto Ms. Aquarian antiquarian, a *Rolling Stone*-caliber journalist, or Alan Greenspan's secret advisor. He had a way of coaxing people into their deeper selves without their knowing it, making them accountable to the gods.

He was sort of goofy too—I picture him on the roof of an ashram during his annual winter sojourn in India, trying to get his Patriots in their first Super Bowl by a half-assed antenna—not long after his pilgrimage to one of the largest gatherings of humanity in history (on the Ganges River, where some of his ashes later floated inexplicably against the current, a final prank, before plunging respectfully downstream). Later he discovered the ashram of yogis who lifted weights with their penises, an experience he delighted in re-telling with mock pious diffidence.

In retrospect, his physical life now abruptly and inexplicably terminated (a fact that doesn't compute as I stare at a snapshot of him smiling disarmingly while paddling a canoe on Webb Pond—an expression of "what am I doing here?" melded with sly earthbound acceptance, a lightness and candor few have), I realize this guy was one of the true clown saints of our time and I find his teaching indisputable and wonder why I didn't shout it back to him, call him on it even once, as my cosmic joke on his: "You avatar, you...!"

Ex-Wilbraham (Mass.) shortstop; Colby College hippie; communal chef who no longer could fry even a veggie burger; thin as a breatharian; guy off the street who somehow convinced an officer at the Bar Harbor Bank to give him a building on Cottage Street, no money down, and turned it into an Asian import business, jewelry workshop, and community squatting ground, while (also) leveraging it into odd properties in Maine, Nova Scotia, Prince Edward Island, and Virginia, capped by a mango-grove, thermal-pool retreat center on the big Hawaiian island; always true to Kirpal Singh and his lineage; who, after the master's passing, became the only disciple to break bread with both embattled branches, those of Ajaib and Darshan Singh, in fact who was recognized and received into the homes of the most inaccessible spiritual teachers and pundits everywhere, this ordinary American kid, because he walked through every door in the most casual, brazen, unadorned, transparent way; daredevil son of Milarepa, who climbed over posted warning signs on Cadillac to mug at the edge of cliffs and then pretend to stumble as if to fall to certain death; who heard the choir of spirits singing continuously at the site of the obliterated World Trade Center and reported that their sacrifice had opened a new portal in the universe for all of us, that they were chanting in ecstatic grief to assist a coming planetary transformation; who read Hindu holy books and sports pages on the steering wheel of his van as he drove to fairs and trade shows as far as D.C. and Detroit, but mostly in the Northeast, rain, sleet, or fog, and of course during one of the worst blizzards in New York State history, to display his home-made rings and pendants; who at a mere beckon would drop what he was doing to join a friend using the perfect clay of a Seal Cove rivulet to assemble Shiva mudpie lingams in the rain and bay mantras at them; who somehow convinced Lindy and me, despite a home and business a continent away, to purchase a house on the island so that we could be his neighbors and pals and participate in the tao of the galactic shopkeeper and enlightened fool.

The character he played in life was so obvious and close to all of us that it was impossible to see him as he was or remember from moment to moment what we were seeing. We were privileged to attend not a saint (god, would he have had a field day over that, mocking and spoofing me—I think maybe the depth of his practice was a secret even to himself, to Richard Handel if not to the cosmic caretaker and witness who wore the body)—we were privileged to be in the company of a human being living as a human being, not making any more karma than he had to, cultivating a dharma so spare and intrinsic he couldn't flaunt it or let it get in the way of R. Handel, regular guy, lazy doofus, best buddy, neo-slacker dude. It is a practice available to all of us. And because this guy lived it and is no longer here, it is bequeathed as a yardstick, a paradigm by which to try to live, a simple yet totally incredible thing.

He has left behind the final word, not in a bad, not in a bad way at all; in fact in a really good way that calls attention to the fact that we all can be kind and gracious without affect of spirituality, without false modesty, without credentials, without fabrication.

The urgency of his passing is to try myself to be quieter, lighter, more honest, less selfish, less narcissistic, less impressed by the things I have done—because achievements in that sense don't stand up to the mountains or the stream of worlds in the Milky Way, the naked truth of empty practice as demonstrated by Mr. Handel.

I would want to honor him by not letting who he was in this community die, by trying to do the things he would do (those *bodhichittas* sometimes called "random acts of kindness"), so that years from now when he is a memory that we have to recall (was he really here, among us; did we really know him; did he pull this off so artfully in our midst?), his unhindered spirit will have suffused us, and we will have become him. Which, after all, is no big deal. Just treat others like kings, like children of God. And don't take yourself too seriously. That's practice enough. That's lesson enough to leave behind in a strangely savage death from which he departed like a feather floating up.

In the spring of '04 Chris Kaiser, instructed by his spirit guides, began a monastic fast. Three weeks into purification without food or water, he fell off his bicycle, smacking his head. His room-mate drove him to the hospital, where he insisted on a swift release—the stay was interrupting the ceremony. Stubborn and temerarious to the end, he told his friends, "You don't see them, but I do." He meant to reassure them that he was under supervision, that no one need worry.

He was still purifying when he mounted his bike the next day. He vanished for the better part of a week, though search parties canvassed the Park and environs. Wendell walked Acadia Mountain and Saint Sauveur with his dogs; Aubrey hailed him while crossing the same terrain at another angle; others tracked futilely from Somes Sound north.

Chris' unrecognizable, emaciated body was found floating in Hamilton Pond by Salisbury Cove, a week or so after he drowned, probably from heart or kidney failure while swimming, the consequence of extreme fasting. Hamilton Pond was where, he wrote, he had located "the vortex to the other world." He swam clothed and bearing ropes, he explained, so that the extra weight might make his marathon more strenuous, hence more pleasing to the gods.

On the night of Chris' actual death, Wendell reported that his dog Boomer, who always positioned himself to guard the master's bed, jumped up as if to defend against an intruder but then began to whine miserably, as though in terror of something more powerful than he could grok. After that he catapulted through the door, knocked over a lamp in the living room, and tried to crawl under a chair that was too small for him. "No doubt about it," Wendell said; "Chris had come to me to say his final good-byes."

Now that he is gone, there is no other possibility except to acknowledge that he made it through the vortex. He saw the light of the other world and swam for it. Anything else would trivialize his act and dishonor his memory. In sacrificing himself, he rescued his text from take-it-or-leave-it New Age frivolity. He believed so much in the reality of sacred sites and spirits that he risked everything on their behalf. He cared enough about the chakra system of Mount Desert Island to give his life for it. No one will read his book the same way again. His pantheon will live on beyond him and become more and more real each year that passes, for every neophyte who visits the island looking for his power spots and shrines, for every debunker who dismissed man-mountain teaching clumsy t'ai chi outside the Bar Harbor YMCA.

Once the most visible presence on Mount Desert Island, Chris Kaiser can't be seen there at all. He has crossed the bar and been taken by the greater tide.

The year was 1963. I was nineteen years old, a sophomore at Amherst College in Massachusetts.* Unknown to me, I had just met the woman I was going to marry. Lindy was a sophomore at Smith, dating an architect at Penn. Neither of us could have imagined the future, even a week beyond, for John

*Some portions of this story are rewritten from my memoir *New Moon;* one small section comes from my memoir *Out of Babylon.* The 1970 episode is reconstructed from an unpublished book called *Episodes in Disguise of a Marriage.* I have changed a few names to protect privacy. I wrote this initially for *The Sun.* The editor and I had discussed my doing a series of memoir pieces as a regular feature of the magazine; yet, commenting on this first story, he said he "didn't get it." Then he lobbied for Schuy's turning out to be gay; otherwise, he remarked, "there was no point." If he thought "being gay" would have been more stunning or decisive than what actually happened, it is evident why my career as *Sun* contributor ended before it began.

F. Kennedy had given a speech at Amherst and was headed to Texas.

I was barely removed from a troubled adolescence—the term "dysfunctional family" was not yet in vogue—and six years of hibernation in an all-boys' private school in New York City. Amherst was all-male also back then.

Schuyler and I were both members of the class of '66, loners though in different ways. I was a wannabe writer with minimal social skills, in fact seriously in danger of being an asshole. I was solemn, tragic, romantic, neurotic, intellectual—also an urban Jew in a WASP culture (and utterly blind to the consequences). Freshman year I made all the wrong moves from the start and irritated the jocks on the fourth floor of James Hall. One night, six weeks into my first term, I was locked in my room (a clever saboteur had rigged the knob so that it would come off in my hand as I closed the door behind me); then lighter fluid was squirted under the handleless barrier and matches struck, setting fires I kept stamping out. The event culminated with me leaning from my window and threshing a hockey stick back and forth, smashing windows within reach, while my tormentors, heads poking in and out those windows, chanted for me to jump. Then two good samaritans from down the hall wrestled through the crowd; they pulled away the ones pounding on my door and, enforcing restraint, ended the episode.

By contrast, Schuyler Pardee was a glamorous James Dean castaway with a heroic snarl and a disaffected lope. A vintage rebel without a cause, he had somehow matriculated through a prominent New England boarding school.

I was carrying a hockey stick again, this time during an intramural match, when Schuyler's path and mine crossed fatefully. By rink rule you weren't allowed to raise the puck, but a pass of mine whizzed by his shins. "Do that again," he called out, brandishing his stick like a cutlass, "and I'll castrate you."

His remark had the curious effect of melting the distance between us and, heading from our lockers to the quad, we fell automatically into each other's company and began dissecting the teacher of our one mutual class, the noted American Studies seer Leo Marx. That I was one of Marx's favorites Schuy considered a *de facto* betrayal because Marx had made him his whipping boy—chief target for pedagogical sarcasm.

Marx liked to foment discussion by playing provocateur. "I guess I'm just too stupid," Schuy retorted on more than one occasion, as the professor baited him with various ruses of cultural interpretation. During another exchange, the whipping boy crooned, "You're smarter and better educated than this here poor dumb plantation slave." Schuy tossed an arsenal of comebacks like these at Marx with a kind of defiant insouciance that earned him increasing scorn and unjustly low grades.

I assured Schuy that the famous professor was now down on me too, since I had been trumpeting his enemies (avant-garde artists like Charles Olson and Stan Brakhage, extracurricular philosopher Carl Jung, *et al.*), and living in renegade, disbarred Phi Psi among the school's radical fringe, a cadre despised by Marx (despite his public liberality). [In Amherst of the mid-sixties, most upperclassmen lived in fraternities because there was a scarcity of other dormitory space, but only Phi Psi, cut loose by its national for admitting a black in the '50s, functioned as a "non-fraternity" fraternity, a nesting site for bikers, musicians, politicos, poets, and science nerds.]

"He's supposed to be one of the good guys," Schuy explained, as we explored our new-found camaraderie that afternoon on the dinner line, "but he's just another authoritarian 'my word goes' *bwana*. He's all for revolution and protest, but in his own class he can't tolerate someone else's opinion without losing his cool. Just because he's teaching *Growing Up Absurd* doesn't mean he's on our side."

In fact, a few months later the author of that book, Paul Goodman, visited our class (he was a friend of Marx). A group of us dined with him afterwards. I had expected a thoughtful, compassionate elder, so was unprepared for the vulgar egoist who disdained a heartfelt question about alienation by saying, "Don't bore me. Kids your age care about nothing but fucking."

As Goodman continued to preen from a mantle of smug machohood, Schuy finally injected, "You can't corrupt us, so why don't you keep your shit to yourself." The table fell under stunned silence, Marx staring death rays. Then Schuy elbowed me, and we departed inelegantly together. As long as twenty years later Marx was still selling his version of my betrayal to a new generation of students.

As I got to know Schuy better over the following months, I realized how much of a paradox he was. A prep-school icon in appearance and style, he was as bright as anyone I knew, also a great natural athlete (though not on any team), and strikingly handsome—a muss of sandy hair, classic patrician features, Kennedy rascality. Yet he was alienated from every peer group at Amherst, including mine. "Just another social club," he retorted when I tried to get him to hang out at Phi Psi, "a lot of ignorant guys practicing reverse snobbism."

He was not trying to date girls, though he was very into them. He took my unabashed enthusiasm for the social events called "mixers" as a sign of weakness, and he urged me to ignore the whole rigmarole because it was a set-up. The entire

adult world was a sham to Schuy, a Madison Avenue invention of fake slick men and Barbie Doll women. He was intend-
ing to stay a pure, untainted lad. "We're enslaved by these
advertising images," he railed after we attended a film showing
by Brakhage, "all the crap that we're supposed to be so that
we can look like soapsuds men, so girls will like us, so we can
get jobs. Brakhage is outside of that, so he's able to make his
own things—and without the derivative academic bullshit of
Marx and his buddies."

Guided by his sense of being hustled by just about everything, Schuy was determined not to be taken in if he could help
it. One Saturday night, however, he dropped his scruples and
we hitched to a mixer at Smith, talking excitement all the way—
his acerbic social commentary, my innocent hopefulness.

Neither of us dared conversation with an actual girl. Standing in the thicket of males outnumbering females by at least
ten to one, we grumbled to each other as more courageous col-
legians sought conversation and collected precious names and
numbers. While I was paralyzed and frustrated, Schuy was
noble and aloof, providing a volley of churlish remarks. I thought
he was far and away the most attractive guy in the room.

"They're so damn good-looking," he sighed, "but that's
what they're playing at. Let one of them act like a normal
human being and come over and ask *me* to dance."

I knew Schuy for only two and a half years. During that time
we had many adventures together, and I absolutely adored him,
not from some lingering boy attraction but as the epitome of the
brave and cool and swashbuckling; he modeled a transition to
a different, more authentic masculinity, outside Ivy League
suavity. I never mastered his insolent, fake-modest shrug of

"sorry, guess I'm no damn good," though I tried often enough that some of my other friends took to spoofing my Pardee imitations. I was too earnest to pull off something so cynical and cool.

Schuy portrayed tragicness well, with spunk and dignity; I bumbled through it with introspective self-pity. It was as though he were in training for a secret mission on which all our futures depended (in fact, he initiated me into *Lord of the Rings* well before the general populace discovered the story). He was Strider, Aragorn; I, at best, a hobbit.

When I have fantasized a movie made from my memoir *New Moon,* it has been Matt Damon in the role of Schuy, although it could have been the other way around. With time and destiny reversed, Schuyler Pardee would have been a more beautiful Matt Damon, sullen, charmingly imperious, the uncorrupted man-child as movie star.

That spring Lindy and I began to go out for real, an on-again, off-again courtship that ran from the middle of sophomore year until our decision to throw in together the fall of senior year. During that same term Schuy became buddies with The Lund Man and Koscis, two disaffected soulmates in our Lawrence seminar. The four of us held running satire during lunch after class, as we spoofed Amherst's mainstream fashions. Sitting in Valentine dining hall, we pointed out examples of "big man," "cowboy cool," "gym warrior," and other pseudointellectual wise guys and pricks. L and K were rogue cowboys of their own ilk. Decked out in jeans and second-hand leather, they shared an old, souped-up sedan and spoke in periodic Laurentian mime.

One Saturday Lund Man drove us all to a swimming hole

up near Vermont: Koscis, Schuy, me, and our dates (Schuy
had a brief flirtation with a Smith girl who published sexy
poems under the name "Wendy Bitter"). I felt an ancient wist-
fulness, as Lindy and I lay on our backs in the grass ... clouds
blown apart in the jetstream. I was chasing the bare eclipse
of a form, itself a shadow. Beyond the hill, the land dipped
precipitously into the unknown, an obliquity that masked a
dream. Something indelible was lost; something equally remote
still beckoned.

Koscis was our hero, king of "guerrilla warfare," conducted
in the Psi U basement on Saturdays after midnight. "It's beyond
description," Schuy warned. "If you could see it you'd realize
that Phi Psi is a bunch of wimps."

In May he extended a guarded invitation: Koscis had arranged
for, of all people, "moi," to witness a session of war as a non-
combatant. He couldn't a hundred percent ensure my safety,
but I would be under his protection. If he were defeated, though,
it was every man for himself.

On the appointed eve with a whispered password, Schuy
led me past a guard, down into the cellar crammed wall to wall
with bodies and shadows, a single lantern hardly illuminating.
Occasionally someone let out a shout.

Suddenly—with a parody of a blood-curdling scream—
Koscis leaped from nowhere onto the bar. His chest painted in
blocks and bands of color, he stared down the group, then
danced in place as others threw objects at him—mostly their
cups of beer, water balloons, light bulbs. He retaliated with
the hose from the keg.

The repartee became more frenzied and, as competitors tried
to drag the king down, he goaded them with obscene taunts

and gestures. Soon a full-scale skirmish broke out. Schuy whispered, "Stay close ... watch. Do you see *The Plumed Serpent?*"

He meant Koscis's favorite Lawrence novel, in which males transcend their mediocre social condition and enact soul-magic. The ritual at the bar was Psi U's attempt at primitive courage, a way of striking back against the allure of women.

As some of the brethren lit hand-made torches, the room shimmered. Koscis demolished the keg with a hammer; the Lund Man applied a torch. Gyrating before the fire, twirling a lance, dislodging challengers bearing sticks and ropes, K had become a parody of late baroque Lawrence, fascist yet primordial. [I read now that he is a renowned psychiatrist.]

A few weeks later at my invitation, Schuy joined my Abnormal Psych class for the movie *David and Lisa*. Lisa was this mute, schizophrenic girl-child, darkly beautiful, played by Janet Margolin; David (Keir Dullea) was an uptight compulsive teenager, obsessed with clocks and death, phobic about being touched. They were residents of the same mental hospital and, gradually through the story, drew each other out by empathy and halting friendship. In the culminative scene Lisa walked up to David, her hand extended ... and he finally allowed her to touch him.

While I was enchanted, Schuy was enraged. "What was wrong with David anyway?" he demanded. "He didn't want to be touched. He understood clocks were the enemy, that they led only to death. Why couldn't they leave him alone? Why did some pretty girl have to come along and invade him?"

That summer Lindy returned to her native Denver and took up with an old boyfriend. Living at my father's hotel in the

Catskills and working for a county newspaper, I felt both heart-broken and intrepid. A sense of the irreplaceability of love would not set in for many years, so I was going out with a foxy waitress named Jean, trying to transfer my fate and affections to her. Meanwhile Schuy had followed his family to Martha's Vineyard where he was working as a dishwasher and racing sailboats. He wrote:

"It sounds to me like the best thing about this Jean is that this all gives you a chance to get Lindy in some kind of per-spective—get some of your power back, speaking 'David-wise'—which turns out to be the same problem I have here. The first day on the job I saw this interesting and thin attrac-tive girl—after a day made a remark to her about how the people around seemed all to be so affected by the bureaucracy and so forth, took her out to coffee, etc. All was nice—she turned out to be a Smith grad who writes poetry, hated Smith, didn't date Amherst, is earning money to go to Europe, and anyway I really like her.

"Her name is Diana, like in the song, 'Oh, please, stay by me'—you know, the embarrassing mushy one. I have been dat-ing her a lot. I can't stand her being a waitress right there and me washing dishes. I don't understand what my position with her is, and what I'm trying to do is change her mind about her being 22 and graduated and me being 21 and (I lied to her) a senior. Anyway, I'm trying to act tough—you know, the way Koscis does—to try (I guess) to shake her up. But I'm pretty weak about it. I guess I've been seeing her quite a bit. A week ago we had this big moment at a party, and it was 'I do love you, but I'm not in love with you.'

"I have been sailing every day now. Lund Man is my crew, and we are living together in this little sort of shed-garage apartment. In the first race I did well till we got lost in the fog and had to be towed in."

Coaxing a few days off work, I drove the New York Thruway to the Mass Pike and took a ferry to the Island. Lund met me at the dock and we headed straight for the Seafood Shanty. When I saw brown-haired, sun-tanned Diana—her 1940s style purveyed with a sassy banter—I too was bewitched. She personified a sophisticated femininity that seemed beyond me, though it had long filled my consciousness and dream life. Schuy eventually poked out of the kitchen in his rubber apron to greet me clownishly. He had started a mustache.

The next afternoon while my friends raced, Diana and I sat on the beach trying to spot the faint likeness of their sails. Knowing I was a writer, she had brought along her black binder of poems and read to me. She invoked a landscape of primeval summers, toy boats, and shiny pebbles, her last poem suggesting to a lover that they spend their lives scraping off the insides of Oreo cookies with their teeth. I had brought my tarot, so I set her fortune on the sand.

Later, Schuy showed that he suspected something—if not my admiration of his woman, my collaboration with her as a fellow writer. He had had a dispute with The Lund Man over misunderstood instructions on board, and his mood was growing blacker by the minute. "You two think everything's a joke," he snapped at us. "Well, it's not. It's life and death now." I could not meet his warrior pose, but I was in awe of it, afraid that one day I would be called to do the same.

He smashed the copy of Paul Anka singing "Diana" that I brought him as a gift, saying, "That's exactly the kind of sentimentality that poisons relationships." He was not amused by my suggestion that the song's words could be reversed from *"You're so young, and I'm so old"* to *"I'm so young and . . ."* Later he added, "I have to break her, like a horse"—a line he got either from Koscis or Lawrence or both or maybe Henry Miller. In the Seafood Shanty he had picked up the nickname Scotty, which he loved because it was sailor and lower class and allowed him to be someone else.

He sulkily avoided both of us the next day, though at least he loaned me his car while he worked, and Diana and I drove the single road to the cliffs at the end of the island. Elegant and charismatic, she pointed out sights as we conversed alternately about literature and our friend. "His tough act is really quite silly," she said. "He's such a child." I hardly knew who I was anymore: another hapless suitor? her confidante? my friend's defender or betrayer?

That evening Schuy and Lund Man got into a perverse scuffle, leading to his kicking Lund out: "Go find other lodgings, man." When I tried gratuitously to intercede, he cursed, "Just fucking leave too. I don't want you around either. I'm in a desperate state, but way ahead of you, and you're both in my way." The Lund Man drove me to the dock where I caught the last ferry.

I didn't see Schuy much during junior year. I was totally occupied with Lindy, as she and I traveled to either Boston or New York almost every weekend—museums, poetry readings, experimental films, friends at Bard College, my father's hotel. We became lovers, broke up a few months later and, after weeks of melodrama, got back together. Compared to the seductive apparition of Diana, this was all too essential and real. As the Stones would soon put it, more or less, *"You don't always get what you want, but you do always get what you need."* Lindy was life.

I occasionally would pass Scotty on campus. Dressed like a gentleman, his mustache now full, he was grim and militaristic. We'd exchange guarded words like two pilgrims acknowledging the shadow of Mordor, the epic in which we were pawns. He had reinvented himself as a nose-to-the-grindstone student, a fledgling male preparing for graduate school and a career in educational psychology. He wanted to marry Diana who, having decided not to go to Europe, was working as a legal secretary in an office in New York. Schuy visited her on weekends whenever he could. Like me, he was trying to leap the abyss of adolescence in a single act of purity and courage.

That summer Lindy and I lived together in a cabin in the woods outside Aspen and worked odd jobs. Shivering out of bed at daybreak to light the woodstove and fill the kettle with water, we made oatmeal and toast and then sat together, cups of coffee in the high grass and dew. Across the road the Roaring Fork milled pebbles. Our kitten Frodo ran from tree to tree, grasping the white bark with her claws and pulling herself like a lemur.

This was the summer at the absolute center of time. All my puppet selves stopped their frenetic imitation of life. I learned what it was like to dwell on the Earth in peace, to feel existence as simple as a thread flowing from a spool. I was no longer at war with the fact or frame of my being, or the myth of my traumatic past.

Schuy and Diana eloped that July, leaving their parents out of it, because, as Schuy put it, "My mother was actually going to thank her for marrying me." They went back to Martha's

Vineyard from where Diana posted this chatty letter: "I'm wait-ressing at the Seafood Shanty, and Schuy started this morning as a short-order cook in the dairy bar right next door. We've really had an existence of sea urchins, growing brown on the rocky beach that is all ours, dipping into the huge lagoon when the sweat gathers in creases of knee and elbow. We are read-ing *The Sound and the Fury* to each other, and Schuy's even tossing out some Pound riddles while I sweep the floor or cook."

In late August while Lindy stayed in Denver with her parents, I drove back East. In the Indian summer before school she took up with her old Penn boyfriend. The word was "devastated": that October I walked around in gloom and despair, unwill-ing even to go to class. Schuy and Diana, having rented a cabin on a lake in Belchertown, became my solace, and I drove reg-ularly to the sanctuary of their relationship. "Don't kid your-self," Diana confided, as we sat by a pond littered with leaves. "Scotty said he was going to break me like a horse, and I'll tell you, I feel very much like a pony that's been hitched. We've got all of that to work through now, and I doubt that we'll make it."

I didn't hear; I saw only a girl and a guy in dungarees and blue work-shirts kicking a soccer ball back and forth in the sun; their situation was idyllic.

One night I completely lost it, wild and crazy, in and out of panic, threatening to jump off the Phi Psi roof (though I wasn't serious). A friend called Schuy.

He came with Diana, and they brought me to their place where I slept on a mattress under a collection of blankets and quilts that they threw on top of me. Rain and branches blew against windows and wall. I drifted through fragments of dreams, awaking each time with a start. I didn't want daylight to come, the birds' chirping to begin. The echo in my mind of *"Try to remember a time in September ..."* * made life on Earth seem irreparably brittle and heart-rending, profound beyond my capacity to bear. Could I ever have been *"a tender and callow fellow"?*

No ... clearly not, but the words themselves were like a dirge, a dirge bearing inside its lyrics and melody a summons to somewhere safe and unknown yet familiar, a sanctum all around me that I could not quite enter, a child I knew intimately but had never been, a man I didn't know how to become, a promise that could not, in the end, be broken. Its tune was a reprieve straight out of the great and beautiful universe, the omen of my own existence in all its wonder and mystery and sustaining texture reaching to me; but I couldn't get at it— so haunting, like thwarted love, like gallant impossible hope. That unknown tender and callow fellow moved me to tears.

Then I heard arguing, its pitch increasing, so I quickly got dressed and went outside.

It was all pain, in all directions pain. There wasn't a clue anywhere. The natural world seemed to stretch unabated in every direction, forever. This was the heart of Mordor.

"It's not so good where we are either," Diana inserted contrarily while Schuy was showering. "We're talking about splitting up." I didn't believe her; I thought it was just grown-up shoptalk. At breakfast they jousted black humor about their upcoming divorce. Diana said she was going to find the Afro

---

*The Fantasticks, book and lyrics by Tom Jones, music by Harvey Schmidt, 1960.

artist she used to flirt with in New York. Schuy talked about a
girl who wore a miniskirt to class, and invoked Wendy Bitter.

I told them I thought their life was beautiful, even their arguing was beautiful. I warned them there wouldn't be anything
better. They smiled at my naiveté; then Schuy drove me back to
Amherst on his way to class.

That was the last time I ever saw him, except maybe at a distance on campus. Maybe. He didn't come to graduation.

Lindy and I were married and living in Ann Arbor two and
a half years later when I read in the *Amherst Magazine* (fall,
1968) that Schuyler Pardee had been killed in a motorcycle
accident in Illinois. I located Diana through Smith College:

"You have no idea how good it is to hear from you. Schuy
left in August after senior year for Southern Illinois University,
having decided he needed some time on his own to get himself
together. Our marriage that spring and summer was, as you
know, very rough. The details—which I would certainly not
be mysterious about, given the right time and place—are simply too complex, and I can't manage them tonight. I can say,
tho', that I never wanted to be away from Schuy, would never
have been separated from him. The divorce was for me the only
way to start over again—toward a new life with Schuy or with
anyone else. I wanted to be with Schuy but jesus christ I couldn't
DO anything, and he seldom if ever acknowledged my letters.
The last I ever saw him was Christmas two years ago. I didn't
know I would never see him again. He was killed this summer,
in June, crossing a bridge. I'm told he was extremely depressed,
had several incompletes in his courses, and had been fooling
around with drugs (which I knew from his visit and worried a
great deal about). It would seem he was entirely out of place in

that conservative southern midwestern school. He was teaching and counseling undergraduates, and they did not take to his independent ideas of education, radical stances with regard to the University and U.S. policy, and ways of dressing. In June they made it known that his services 'would not be needed.'

"It's pretty clear, I think, how brutal this was. I feel he gave up any interest in staying alive. He bought the cycle that spring and often took incredible chances on it. In a diary he wrote that man was ruled by fate and for him darkly so. That was not the Schuy I knew. But I met and talked with one person he knew well and have heard of others who were touched by that beautiful thing Schuy was. So I know it was not just a lovely dream I had.

"It was killing to lose him once. Impossible to lose him forever. I'm so grateful for the two years I had with him."

August, 1970: Lindy and I were living in Cape Elizabeth, Maine, with our year-old son. After a year of ethnography with fishermen on Mount Desert Island, I was about to start teaching at the University in Portland. One afternoon we came home to a message on our machine from Diana, saying that she was visiting friends nearby and hoping to come see us. Earlier that summer, I had inscribed a copy of my first book, *Solar Journal,* "For Diana, whom I continue to honor as a poet and a woman," and sent it to her New York address.

It was a different era then, more innocent, idealistic, socially experimental. I won't describe how it came to pass (that's a different story), but Diana and I set out on a day together down the coast. We didn't know how it would turn out but, to my mind, it was a romantic tryst.

"What made you think," she asked, upon finally dragging

my fantasy out of me, "that I'd agree to something like this?"

I didn't know what we had agreed to but, faced with her ire, conceded we were merely on a short drive and would come back after lunch.

I directed her toward Highway 1 and then was quiet, feeling only the sunlight, the seductive strangeness of her car, her driving. I wanted to take it all in—in case there was nothing else—the smell of the interior, the glance out the corner of my eye at her forbidden body, the eyes-on-the-road stare of Schuy's wife. I was shivering as though it were cold; in fact, it was.

She described Schuy's last adventure. He had pressed his girlfriend's hand against his own, whispering, "Don't leave me." As he gunned the vehicle, a curve came on, a bridge over a river. His body ended up in the stream below. She was found wandering in tiny circles. Released from the hospital, she went straight to the abandoned wife to bear witness.

Winding along the coast highway, Diana had other stories, most of which I forget, though I remember one about when she and Schuy were living in Belchertown and an Amherst professor, the most prominent of the tenured literati there—an author of tales of marital infidelities and also Schuy's advisor—had propositioned her.

"I was never more flattered in my life. I said no only for Schuy's sake."

I felt the old allure of their marriage, fueled additionally by fancies of revenge: this powerful male was Schuy's mentor and the first of the so-called "bad guys," the self-satisfied literary pricks, that he had actually trusted. What right did *he* have betraying his student—my friend?

My outrage did not entirely eclipse the counter-innuendo: What right did I...?

Hunger came after two hours in the salt air at Rockland. We sat on benches of a Dairy Queen and watched boats (as we had years earlier on Martha's Vineyard). Diana kneeled by the bushes and coaxed a kitten. It sat on her lap and then climbed onto the table. I didn't have the cards with me, but telling her fortune no longer required them.

On the journey back—the local star bending into one of its most arcane sunsets—I expressed the wonder that we were any place at all, our chariot moving, light shifting alchemically purple into dusk, the harbingers of constellations. It seemed so unreal—Schuy no longer alive; me married with a son; yet driving in her rebel car from nowhere to nowhere. I felt his ambiance so strongly, as though she were a form of him. His old desire for this woman met my desire for her now, and my body felt his bravado inside it.

"We could be on another planet; these lives we are given," I said, "are strange beyond belief." In that spell, there was suddenly an Athena figure piloting her vehicle who uttered impossible words: "Yes, let's. . . ." They rang through layers of my understanding, but not understanding—never understanding—all the way to eros, which couldn't believe it, that its most secret fantasy would be fulfilled.

She pulled into a cluster of bungalows and parked by the main cottage. "I'll handle this," she said, hopping out. The song I heard in my head had both everything and nothing to do with it: *"Her name was Joanne, and she lived in a meadow lying. . . ."*

She showily dangled the key in front of me (a playful Diana I hadn't known . . .). She had also procured a six-pack. I got out of the car and we walked to the door. It was a cottage among trees along the Sheepscott River. Inside was stone cold.

She set the thermostat; we stood by the electric wall blower wearing our coats, beer cans in hand, sipping. I listened to her voice that had borrowed so much of Schuy's idiom that she was both of them now. I finally put my arm around her back and felt her solidity, her long legs stretching out. She had such a pensive intense gaze. As it was getting warm, we lay down alongside each other and embraced.

Now I was part of their life together, inside the fraternity of Diana and Schuy; only it was my body instead of his, my body swimming upstream toward her like some unquenchable salmon, not as perfect or beautiful, not as charismatic. After all, he was Adonis, JFK; he was *On the Waterfront.* She was his forever co-star—June Allyson opposite Peter Lawford in *Good News,* the sweetheart of Sigma Chi, the heroine of *The Summer of '42.*

Who was I? Insubstantial, gangly, dark, problematic, an outsider male. But at least I was alive, and wanted her.

We explored our infatuation with Schuy unhindered, rein-vented his pluck and beauty, took license with his own inscrutable yearning for us, and finally brought him back to life by our care for who he was in our hunger for each other's bodies, swallowing his incalculable existence into our own, in awe and in love, and in the irrevocability of desire. Not only did I want her; I was grateful to be in this world, to still be real.

I imagined she reached for me as not only his friend but his antipode, as the father of a child she might have had ("he's so immaculate," she remarked of my son; "he smells so clean and new"), also in an act of sacred disobedience and resurrection.

I had overwhelming compassion for her, her brittleness and underbelly of hardened grief. Anything I could have given of me I would have, for just that one night she took care of me in Belchertown when I thought I wouldn't live. She moseyed her hand across my face, said mysteriously, "Enough!" … leaned down, and put her tongue so deep in me I poured whatever was left into her too. And remembered the rest of the words: "*… broke down her desires/like the light through a prism … /into yellow and blues/and a tune which I could not have sung.*"

We stood outside in moonlight while a young couple with a child checked into another cottage, looking like her and me, like Lindy and me, like Schuy and her.

I thought, 'It will be a mystery forever. I will never touch this one.'

I saw Diana again briefly a few months later at a reading I gave in New Haven. After that we exchanged letters for a year or so.

In 1972 Lindy and I moved to Vermont, and five years hence we transported our life to California in two cars and a U-Haul. Our initiation there was jobless, sustained by freelance and part-time work. In fact, Diana's unrequited professor-suitor singlehandedly blocked a couple of promising grants for me— I wanted to continue my ethnographic research with a comparative study of Hopi farmers and Tlingit fishermen. I had never taken a course with the man, and he had never met me (nor, of course, was he an anthropologist), but among the higher echelons of Amherst it was routine to disapprove of whatever I did.

I know this sounds almost conspiratorial, but to this day, approaching my forty-year reunion, I feel the enigma of an Ivy League college that would *never* banish any of its sons formally but tries still to vomit out those who oppose its patriotic Western paradigms and corporate allegiances. That's why Schuy was so down on them. He knew they were our enemies. They would kill us just for sport.

Eventually Lindy's and my tiny publishing company, North Atlantic Books, specializing initially in t'ai chi, homeopathy, and baseball literature, turned into a full-fledged business with real employees. By 1991 our son had grown up and gone to college. Our daughter, born in Vermont, had become a Berkeley teenager. I was studying craniosacral therapy, learning to track pulses and energies in tissues. That year, Lindy and I attended our twenty-fifth college reunions at Smith and Amherst.

At the Saturday banquet at Amherst, a time was set aside for speeches about deceased classmates. Half a dozen members wanted to reminisce about Marshall Bloom, founder of the Liberation News Service, a countercultural superstar and the hero of Ray Mungo's book *Famous Long Ago,* who committed suicide by carbon monoxide in a garage on a nearby Massachusetts commune about the time Schuyler rode his bike off a Carbondale bridge. Marshall was struggling with coming out gay, about a decade and a half before the world was ready for it.

While Marshall was also a friend of mine, I was the only one in our class who really knew Schuy, so I alone recalled his sassiness, quirky sense of honor, outrageous spirit. Afterwards William Pritchard, an English professor who attended our banquet because he spanned the decades, asked me if I knew that Schuyler's ex-wife had come back to Amherst, married a colleague of his (not the one who had propositioned her), later gotten divorced, and was now in Belchertown with a lesbian partner. I was astonished. She had gone flesh and blood into the English department that Schuy and I despised—and was now out of it and in another life.

A year later, while taking our daughter to look at colleges in New England, I tracked Diana down for dinner.

We all age in different ways at different speeds. Diana had been older than Schuy and me, but only by a couple of years. Yet the woman who opened the door that night was inconceivable; she had passed into an indeterminate, haggard old age. The elegance, sandy beauty, regality, and spunky dance were gone. After a few moments of conversation I realized that her sweetness and droll wit had metastasized into venom. She had become a sacred crone, the numinous dark side of the Great Mother. Worn down by relational disappointments, health setbacks, and poverty, she was fierce and acrimonious, and wasted little time demolishing me. She chided me for our indiscreet fling. She put down my writing as undisciplined and indulgent, implying that she had been remiss in not reporting this years earlier. She attacked North Atlantic Books as vanity publishing: I was a fraud, inventing my own private universe because I could not live up to the academic standards by which we had all been trained.

It only got worse during dinner; at various times she accused me of being gullible, self-involved, narcissistic, whimsical, out of touch with the real world. She didn't want to talk about Schuy. She was angry at him too, for his fabricated macho games and casual discarding of her—"like kleenex," she said. He was someone to whom she was married very briefly a long time ago. That's all. The waters had closed over him, in fact over our entire youth. The era that made Schuy charming was now as lost to us as the milieu of the Charleston and rumble-seat was to bewildered alumni strolling across our Amherst campus of the '60s.

On the way back to the car Diana said, "Here's a story that will probably interest you," implying it was of no significance otherwise: "A friend gave me a gift certificate as a birthday present last year—it was a visit to a psychic, the kind of thing *you* would like. I don't believe in that sort of stuff, but I didn't want to insult her, so I went. The moment I sat down, the lady told me, 'I have a message for you from your husband.' I assumed she meant Basil, but she said, 'No, your other husband, the one who rode the motorcycle. He wants you to know: he's happier without a body.'"

I have had many experiences with psychics and, more to the point, for the last twenty-five years have been hanging out with people who discuss spirit conversation, other dimensions, remote viewing, and being in and out of bodies as naturally as most banter politics and weather—so I am not bowled over by run-of-the-mill channeling. Yet this message from "Schuy" meant something I couldn't quite grasp. Through the rest of the '90s, across Y2K, I kept returning to it in my mind. It became a part of my truth about him, and myself.

Recognition came ten and a half years later, the summer of 2003 on Mount Desert Island. I hiked there regularly with my friend Bill Kotzwinkle, spiritual warrior and author of our company's best-selling book. Up and down Western Mountain we discussed literature, philosophy, dreams, Bach flower remedies, the unconscious, and our mortality. Amazed at having become old, we inspired and scared each other with ghost tales, secret paranoias about the universe.

Who deeded us these lives? How can we protect ourselves against an eagle who devours souls? How do we avoid spiritual campfire songs and false gurus and still honor the gods? How do we prepare for our inescapable appointment with the infinite? Bill was concerned to meet death each morning with devotionals and Iron Shirt *chi gung,* whereas I figured—our piety, hard work, and sincerity aside—the universe was going to bounce us where it pleased. Such was our proud dissension, conducted from different angles at many levels.

On one of these hikes, as we were huffing and puffing our way through the damp woods above Seal Cove Pond, exchanging thoughts on the topic of reincarnation—that is, whenever breath permitted speech—I was seized by a sudden and inexplicable compulsion to tell him the story of Schuyler. By comparison to our exquisitely nuanced Tibetan Buddhist and Castanedan excursions it proved a kitsch and overwrought narrative —in fact, I was sure Bill would cut me off as per his "suffer no fools" regime. Yet he was attentive, even spellbound (as I was) by the madcap glee with which I recounted every last drop. Sagas of angelic recital are allotted but a few times in a life, and this was one of mine. As I approached story's end, I felt chills down my spine because I knew I finally had it, the last piece.

I had always figured I would die young or, if not, that everyone I loved would be taken away, making life intolerable— but, no, neither had happened; I had stayed here and aged.

Schuy once had youth and spirit and limitlessness, but he elected to wager these in a throw of dice against death, a heedless gamble that something unendurable in him must have required. Did his false valor come from cosmic grief or an even more profound ennui? How did desire, artless adolescent desire, lead him to the executioner?

"Goddamnit, Bill," I found myself shouting, "he had such a beautiful body. He had style. He was charismatic, a great athlete, bright, irresistible. But he didn't want it. He didn't want his body. But I must have wanted mine. Because here it is, thirty-five years later, and I still have it." Lindy and I were almost sixty; our children had become adults. It was Schuyler who remained fixed and invincible at twenty-three, having left Diana behind in a cabin in Belchertown.

Many years from now, when the consequences of this regime are known, America will be a full-fledged Third World country, a rust-belt mega-slum series of barrios ruled by survivalist armies, motorcycle gangs, and assorted ethnic mafias. It will be Mad Max/Philip K. Dick/tent city, from sea to shining sea. Or think a behemoth multiracial Macedonia. It's already the world's largest banana republic.

India and China will be the overseer civilizations of the next century, commanding the pipelines, using our cheap hillbilly labor to produce their consumer goods. Uruguay and Nigeria will be more prosperous and advanced than the U.S.

Those in power (and those who cheer them on as if America's bounty could burgeon forever) do not understand the laws of sky and rock and food chains, the legendary inexorable ones. Every party lasts only so long, and gluttons and revelers never see in time who among the uninvited is Robespierre, who Charles Taylor, what docile servants melt away into cells of Osama. There are eyes in the darkness watching us more intently than we can imagine: wild cats, stray dogs, rodents, viruses, impoverished wardens of lands we have insolently robbed. There is not an endless supply out there of lackeys, dupes, greenhorns, and migrant workers to exploit. The rest of the world is not going to be pushed around and taken to the cleaners forever. The "next generation" of former customers and vanquished aborigines is developing computer skills, building infrastructures, training warriors, and enriching uranium.

Profligate military adventurism + orchestrated maudlin patriotism + NASCAR Burger King culture + exhaustion of soils and rivers + chemico-industrialization of farmland and Monsantoization of agriculture + increasing oil dependency with squandering at the pump + export of jobs, skills, and industries + hemorrhaging national debt + free-floating yellowbacks at phony interest rates + opportunists selling stocks and real estate back and forth to each other + a work force cloning financial managers, agents, beak-dipping middlemen, day traders, would-be moguls and their personal trainers...drug dealers, pimps, muggers, welfare recipients, con artists, loan sharks, assorted bottom-feeders and hustlers, and the terminally unemployable—with a paucity of real production + a madrasa-like education system overseen by yokels and moral enforcers + capitulation of Congress to lobbyists and regulatory agencies run by industry scabs + coronation of redneck nobilities + neglect of infrastructure, roads, levees, and bridges + corporate executives taking multimillion-dollar bonuses for mediocrity and the mediocrity of the whole system = famine + civil war + a national military composed of addicts and criminals that will make the incipient Iraqi army and child militias of West Africa look spit-and-polish.

The only hope of avoiding this fate, or worse, is to undo the regime's scams and protocols as quickly as possible and ask the world to forgive us for destroying the last best hope of mankind—to become America again, from the proverbial Gulf Stream waters to the redwood forest, from Bangor and Fargo to El Paso and Santa Fe, so long as these remain habitats.

The beetles who have infested this orchard are getting fat and drunk and seriously dug in. Natty Bumpo and Huck Finn better go to work pronto, recruit Dan Boone, Hopalong Cassidy, Abigail Adams, Harriet Tubman, Mother Jones, John Brown, smoke the bastards out of their nests while they're too busy divvying up America the beautiful and laying bumpkin claim to the rest of the world to notice that they are fast running out of trees.

There will be no pity for us in the suburbs of Bombay or Shanghai. In fact, as they watch this embarrassing spectacle of international ineptitude and Three Stooges statesmanship, they wonder why we are so eager to give it all away—our wealth, our superpower status, our military-industrial advantage.

Not so long ago they wanted to be us.

The government here now is, writ large, greasier, more lunkhead and impenitently corrupt, than the worst Jim Crow state house of the 1950s.

Opening the Energy Gates

Matter is thought that has solidified, ideas that have become things. Rock is a concentrated, synopsized scroll of billions of raw untranslated universes. We live in a domain of thought made into mantle and flesh, toads laying low in the muck.

The tiniest hieroglyphic particle contains far more energy than all the fossil fuels pooled for millennia inside the Earth, if only we knew how to tap its riddle and unlock its surge. It doesn't take cubic tons to move engines or civilizations; it takes but a single quark or electron, one snippet of anything. The quest for energy independence, like that for happiness and enlightenment, is located now in precisely the wrong dimension.

The ultimate telekinetic rainbow flashes inertially in the one place we are not looking. By comparison to crude, nonrenewable sources of energy, real energy is unimaginably easy to get at because it comes unbeckoned, unrefined, drop by bottomless, rich drop. However, if you don't know how to summon it or conduct its flow, you can dredge up an entire planet instead.

Computers work miraculously because millions of gigabytes can be stored in a microchip. As electrical engineers are discovering to their delight, there seems no limit to how many texts and pictures can be packed into less and less space. Year by year, as we dig a bigger and bigger hole in the planet, we simultaneously are placing the essential articles of creation on a single flower.

Healing is accomplished solely by conducting energy—the finer the energy, the deeper and more instantaneous the cure. What is holistic originates nowhere and changes everything, by contrast to gross pharms or scalpels that linearly follow an edge or dispersal grid. Pure medicine cures by stirring the primordial breeze that kindles our body-selves, that blows us from potentiality into existence—it activates the breath of life that radiates the latent shape of being as well as the texture and motility of each of the organs.

Transformation comes from impalpably deep inside, flowing outward into a morphogenetic river as if a time-lapse blossom opening. The restorative stream percolates through tissues and cells, from core to periphery, from the completely concealed and internal to the plainly manifest, from energy to matter and back to energy again. Its trajectory can't be tracked or x-rayed because it has already happened, because it is penetrating too fast to see and traveling too slow to measure.

Chi gung is the perfect exercise, for it takes a single image and the breath imbuing it and transmutes it everywhere into viscera and bone. The belief that matter can flat-out change is the key to mind changing it. The arms, sweeping and spiraling above the head, scrape the edges of the entire universe. The vital ball that the palms cup and carry from the belly to the spine up to the third eye is as real as a basketball, if the internal image of it is meticulous and sustained. The crown chakra opens not by force but its opposite, and the chi nectar thereby unleashed travels in the shape of an underlying energy field, through orthodoxies of anatomy and disease, liberating them at the submolecular level. There is no strain to this activity because that would be like trying to budge mountains. Healing is what happens when almost every ounce of strain is relinquished and the tide comes in.

"Séamus O'Brien, Oh Won't You Come Home"

The only possible explanation for what this is about is: it is what we are about. We are expressing an elaborate extended scenery and luminous world drama because it has been our untold story for so long, our own indestructible kernel for perhaps as many as twenty or thirty creations, the sole survivor of the demise of every one of them, camouflaged in all of them in tenebrous darkness, a mere insinuation, a hint of something *else,* ineffable and grand. Once inextricably buried and old, it is now youthful and energetic, being realized here in all its passion and radiance.

The impression that would not go away, yet could never be

located or proven to be anything at all unless a dreaming eclipsed by the unknowable margin of another dream is finally getting lived. And it turns out not to be elusive or vague at all. It is filled with light and opacity and ecstasy and love and wonder and clouds and squirrels and gulls and grieving and horrific violence and epiphany and the Great Dance itself—things that did not manifest in this way or so profoundly in any prior creation.

The only possible explanation for what we are is: we are what this is. Life and death. Being and not being. Dreaming and awakening.

Inside us once, before we were "this"—whatever odd little nubbin of a non-thing we were then, without either landscape or experience—was an intimation of something complicated, vast and colossal beyond conception, a figure or shape shrouded in layers of bottomless and unprobeable desire. Cooped up thus, it was a stale, unprofitable density—a boil that somehow contained the entire cosmos. It had to be lanced, extricated, and disclosed for us to be.

The landscape and the emotional history of the universe are externalized figments of our own nucleus, our long-forgotten essence, like the pea the princess felt through all her mattresses, projected into sky and time. We now wear its meaning as our body and sensations, and we walk in its corridors as reality, space, and event.

The best indication of what is expected of us, of what life is, is that we come into this body and awareness unbidden and uninstructed. We have no choice in the matter, no conscious choice anyway. We matriculate in darkness and wake up here, blind man's bluff. And we must make the best of it, whatever we are and in whatever condition and status we find ourselves. We must do the stuff of our delegated planet, species, time in history, family, inheritance. We must live our characteristics at whatever* degree of sentience and libido we have. If we are a fly, we must behave as a fly. If we are a blue jay, we must race around doing "blue jay" things. If we are a sleepy marsupial or engorged flukeworm, then we must enact those sluggish regimes. If crippled or deaf, we must bumble along happily because the alternative is being someone else or some*thing* else, or nothing at all. If a dwarf or an AIDS baby, we must chivalrously accept our lot. As the saying goes, a pauper or a king.... Which also means a citizen of Somalia or of some tribe on a planet in Andromeda. Either way, ready or not....

And creatures respond very well. They don't immediately complain, ask for a better deal or go on strike. Of course, there is no one to complain to and nothing to strike against except one's self and destiny.

This is our main hope: we got here once upon a time; we made sense of it. Whatever hand we are dealt next, even no hand, we will make the best of, because the only promise is that we will continue to be what we are.

This is the main thing I would tell that sweet little hovering moth, so inconsequential and anonymous, if I could get on

---

*That is the cosmic subtext of the present vernacular "whatever"— a slightly dismissive, indiffferently defiant epistemology for our place in the universe, offered in relation to some inflated thing that inconveniently intrudes its own self-importance or duty. Nothing, under these absurd circumstances, could be that pressing really: "Whatever...."

its tight frequency and transmit through its simple nervous system to its consciousness. I would also tell that spider who stopped briefly on my leg before continuing its path up and down fabric. I would like us to celebrate this condition together and be friends.

But they will have nothing of such sentimentality. They already know it, better than I do, because they don't have to know it. They are too busy carrying out its edict to be bothered by what I am or what plan I have. "Just don't kill me," they think in crude non-thought. "Big shadow, let me pass."

The nature of consciousness is that it springs up like a fire and ignites its own mindedness, everywhere and anywhere. In that sense, all creatures are partners in one Great Dance.

We get in most trouble when we think that we are supposed to have an explicit mission and know what it is. We struggle to stay on course. We put ourselves on timetables. We seek solace against our mortality and seeming plight.

The sole clue of creation is that it exists and we do too. There is nothing more central and hopeful than that. Be what you are with integrity and grace. Accept that you are as tiny and insignificant as a moth. It may seem a curse and a burden at times, but it is also a gift, to get to view and experience all this in some manner. You could never explain it otherwise. It is just too weird and complicated. In fact, you could never, and that would be the real tragedy.

There is no line-up or drill. You get cocooned into some shape or other according to local biological custom and then, here you are, pushed into the day's business, peckish enough to have to get something into your guts, thirsty enough that that rain feels awfully good on your carapace, afraid that what was just given to you, though you hardly requested it or know what it is, will be taken away from you just as peremptorily and unexplained.

In the autumn of their fifth year and for many thereafter, male and female emperor penguins of Antarctica depart the food-teeming waters beneath the ice, plop onto the indelible snow-bank, rise on their saurian hinds, and then shufle in a column seventy miles to their inland breeding grounds.* Clambering up and down hills of snow, sometimes tobogganing along the hardpack, they trek unflaggingly toward the tribal ceremony.

When the survivors reach the ancestral site, their reunion looks like a Woodstock Festival on a glacier. Bumping and screeching, the birds greet raucously and search and compete for mates. When two of them are satisfied with each other, they stand in place, nuzzling and fondling, using their bellies, beaks, and vestigial wings to court. Then they breed monogamously. But they do not immediately return to the ocean after the romancing, for this is where the next generation will be born.

Through two months or so of increasingly frigid, sunless winter, under the waxing and waning Moon and dizzily flash-ing aurora, the flock closes its circle, huddling in a dense macro-organism, pushing closer and closer together against hundred-mile-per-hour winds and temperatures approaching ninety degrees below zero, providing collective body heat and giving off a community sound something like a Tibetan Buddhist chant and something like the hum of converging electrical wires. As eggs gastrulate in the females' bodies and bird homun-

---

*The March of the Penguins, directed by Luc Jacquet, Warner Films, 2005.

culi matriculate, this mantra of incipient penguin language rises and falls atonally, sweeping through vast interludes and exotic sub-themes and fugues in unending segues of semi-harmony, accompanied by the great wind chime and blinding snow. This is the original, floating-continent, pre-Aboriginal, pre-human Dreamtime, as panpipes and didjeridus play through animal bones and lungs and the planet itself.

When a single egg drops from a female, she grasps and balances its ball of energy between her tan-t'ien and reptilian claws and keeps it warm there in a standing yoga posture. Having consumed most of her own yolk in the ovum's manufacture, she must soon transfer its responsibility, delicately and quickly rolling it to her mate who will hold it between his own belly and feet for another two months until it hatches. Famished, she marches seventy miles back to the ice break and plunges into the subglacial sea. As she jets through the ecstatic waters, feeding on the bounty there, the males press even closer together in a circle of feathered, resonating biomass, sounding a lonely dirge of creation, an elegy of life on the planet, as the blizzards intensify and predators circle.

After the chick is born, the male penguin will embrace its fuzzy bundle in his coat against a climate that could freeze a whelp solid in minutes. Maintaining his posture with a birdling in the place of the former egg, though excruciatingly hungry and apparently very cold, vibrating and voicing the great riddle and grief of existence, he will protect the baby until his mate returns, much like those Kiowa warriors whose single role on the battlefield was to pick a position and hold it, weaponless (or with a bow and single un-aimed arrow), against all assaults and intrusions, to cast an aura of ferocity that might render them immune from injury or kill, thereby changing the nature and meaning of the battle, striking if not fear, then recognition, into the hearts of the opponents.

The males do not have to support their asanas forever, for the females, with a banquet of semi-digested fish and krill and jellyfish in their maws, will faithfully honor the species pact and, at great danger and travail to themselves, make the long journey back to nourish the near-starving fledglings. Only then are the males free to reenter the sea and eat.

These are birds, but they cannot fly. Their lovely black, white, and yellow coloring, upright postures, and proximate scale give them the appearance of chubby uniformed humanoids, primates like us. But they are not even close to human. Their calm black slits of eyes show no humanity or mammalian empathy but do suggest some version of love, some cosmic variant of philanthropy and wisdom and benevolence, regalia of a proud but unfamiliar race that seems to belong on another world entirely. They are more closely related to snakes and hawks than to us, but they execute the full semblance of a primitive human village there on the glacier, stand as a lost tribe of shaman bodhisattvas, transformed by bird costumes into flightless apostles who hunt in the coldest waters of this world.

What drives them? What motivates them to leave paradise so bravely and stalwartly? Why do they undertake such a selfless cycle of journeys on which so many must perish?

Just for the preservation of the species? What kind of a reason is that? So much painful labor for such a high risk to reward ratio? After all, despite their rigorous toil, a fair percentage of eggs will roll helplessly from their grasp and freeze to stone. The same fate will befall many of the chicks, while others will be caught and chomped by birds of prey and seals. Serious and on task, the birds don't gripe, puzzle, or dispute. They are in the moment. A penguin may waddle briefly after its mummified egg but then summarily accepts its state.

Is it instinct? Are homing maps built into the penguins' gene mix? If so, how are they constructed enzymatically, where on the helices are they stored, and in what manner are they triggered and projected into sensate activity?

Is it desire? Is it love, or even libido? Or is it something we know nothing about and cannot bind in our philosophy? Is it the irrevocable character of life itself—a signature through which each new creature is incarnated, which each alive thing honors, not objectively from chromosome directive but as the existential, metaphysical fact of being what it is? Is the devoted service of these birds rendered so amiably because they have no other reasonable choice?

Here is how fundamental and deep-seated are the emperor penguins' courtship rituals and altruistic deeds of parenthood: we officially civilized simians cannot advance one iota of humanity beyond them in our courtships, romancing, childbirth, acts of philanthropy, and nurturing of our young. We transform love and kindness and service into self-conscious modalities and render symbols and elaborate dramas out of them; we create whole civilizations based on our vaunted lip service to kinship, morality, charity, and other ideals—things animals know nothing about. We signify and explicate the primordial gift.

Yet we must always return to those same prehuman roots to express love and empathy. We must draw on the same ferocity as Antarctic penguins to survive in nature, achieve our personal identities, and fight our battles and wars. We must stand as courageously and resolutely in place in our own night of mind-created anxieties as emperor penguins do with their eggs in the sheer physical winds without symbols or concepts, and hold our ground in voodoo-ridden crises we have fomented. We and our fellow creatures—all indigenous animals—must change the nature and meaning of the battle. That is the single motive, the one hope, not to think and cobble ourselves into vaunted beings but to act from a reservoir of empathy, selflessness, and mercy that is actually beyond our knowledge and beyond explanation. Only then can we safeguard ourselves, the penguins, and in fact all life on the planet from this second-wave, late Ice Age onslaught of name and commodity. Otherwise, the Poles will melt, and the penguins' habitat and totems will vanish from this clime.

We kid ourselves both that the birds are mindless automatons and we are educated seers, zoo-keepers, and biologists. Instinct is greater than all of our tribal wisdom and education because it is not instinct at all. We see a perfect reflection of selfless, unconditional love in these walking birds because the thing they are teaching is not a sentiment or metaphor or anything we think of these days as love or quarter. They are teaching *fact,* and if we forfeit them as teachers, we will lose fact itself and, shortly thereafter, the thread of love—and then our original nature.

It is now the case that four transnational seed companies control the patents for most of the major grains, fruits, and vegetables of Earth.* This is not even the worst of it. Those four corporations are also collecting and privatizing DNA from the ancient indigenous breeds that make up the Earth's biodiversity

---

*The Future of Food, directed by Deborah Koons Garcia, Lily Films, 2005.

and food supply; they are stealing the legacy of thousands of generations of free farmers on every continent, going back to the Neolithic. These biotech opportunists are grabbing genomes from seed banks like hundred-dollar bills blowing out of an overturned Brinks wagon and registering their own patents for them as if they invented or otherwise earned them. Guess what? We invent nothing. At best we are humble servants.

Some of these species they suppress in thinly disguised genocides. Many they alter in vulgar ways that make the plants less nutritious and ecologically vital but more commercially exploitable. Though claiming via public relations to enhance the agricultural potential of the Earth and feed the masses, they are actually salesmen and racketeers, infiltrating the infrastructure of organisms with bacterial and other foreign agents, switching the chromosomes of plants with not only other plants but animals, solely to make glitzy products and fool the public. Yes, a lawyer with a briefcase can steal more money than a hundred men with guns. Ally that lawyer with self-enamored, contemptuous scientists, and you have the perfect formula for a multibillion-dollar hoax.

Monsanto Company, which manufactures a pesticide called Roundup, also clones Roundup-ready rapeseed (canola) which can be sprayed with virtually unlimited Roundup and survive the attack. That is, insects can be nuked without killing the plants, and who cares about the health of the land or the consumer? The company first makes big money on the pesticide and then even more money from the seed that has to be bought to make use of the pesticide.

But the con is far more insidious than that. As these biotech mongrels escape into the natural world, they hybridize with crops in random farmers' fields as well as with wild varieties of their progenitors. They infest the planetary genome.

And then the caper gets worse. Representatives of these multinational corporations send spies out to check farmers' fields and, when they find their own biotech versions growing in one of them, sue the owners of those fields for violation of their patent. The fact that the farmers have no control over the invasion of their land by these mutants is considered irrelevant to the violation of the law of property. The additional fact that the defendants don't even want the interlopers gives them no exemption, doesn't even accord them the right to countersue.

Courts have repeatedly found in favor of the corporations, handing down decisions explicitly stating that it doesn't matter if GMOs found on private farmland got there by wind, water, or accident; the property owners are responsible for paying the seed companies for their "use." It wouldn't even matter if brigands in the employ of the corporations spread the mutants on purpose. The responsibility for protecting these bogus patents rests with the farmers. Common law and common sense have been turned upside-down in this epidemic of biotech manipulation and cynical abuse.

In addition, crazed engineers at these companies are now working on terminator genes, sequences that will make targeted plants incapable of producing offspring, hence forcing poor farmers throughout the world, who have traditionally saved their seeds for next year, suddenly to have to purchase them anew each season—planned obsolescence transferred from washing machines to organisms. If these terminators get loose, they may hybridize with all manner of domesticated and wild plants, terminating future generations and bringing a silence and famine over the planet.

What idiot bureaucrat would allow the patenting of life in

the first place? What perverse twisting of the meaning of democracy and property rights would lead executives and judiciaries to place the value of short-term personal profit over the long-term value of food and life itself? What fools in robes would rule in favor of corporations over farmers in cases of GMO invasion?

Many disputes in our embattled era have two sides to them, and some things may be mysteries, their verdicts hanging in the balance. But the application of biotech to agriculture and animal husbandry, the invention of botanical and zoological patents and genomic hoarding, the attempt to build new costs and windfall profits into the ancient ceremony of farming, and the defense of this boondoggle by corporations and law firms on retainer are pathology, pure and simple. Not law, not cleverness, not good science, not a misunderstood attempt to end hunger on Earth, not even real business, but stupidity, then greed, then madness and malignancy.

After the lightning struck a transformer in Ellsworth (6:30 in the afternoon, July 27, 2005) we watched the world grow dark by degrees. Mars crystallized over the forest in a coverlet of stars. This is the way it used to happen, day gradually dissolving into night.

Thunder shook the house. From an otherwise empty drawer I salvaged a $5 portable radio that still had a battery, and the sound of "Stand By Me" from some tower to the west and probably the north, filled the candle-lit downstairs with a tinny nostalgia that made the glimmering perfect: *no we won't be afraid....*

The deceptively simple pop lyrics go: "Don't ask me why, but time has passed us by."

I don't mourn the mother who took her life thirty years ago. I mourn the mother I never had—the changeling that never came to this world, the spunky girl who perished before I was born.

I feel nothing but horror for the dark lady who imprisoned us and our childhood in her gulag, the cold vampira my school mates wolf-whistled and eyed as a dish, as she dressed appreciatively to the occasion. I could never imagine feeling anything soft for her, let alone grief, let alone lust.

I don't mourn her charming boy, my paranoid, amok brother who killed himself in Westport last May. I mourn the playmate of rainy New York afternoons and partner in baseball cards and Black Mountain poetry. All my life until his suicide I dreamed of returning to our 96th Street home, only to learn that he had died there when we were children. I would wake astonished and consoled he was still alive, though a ravaged figment of his boyish self.

A grief that seems improperly shallow and vague for such searing tragedy, I have come to understand, is but an evasive and symptomatic semblance of real grief far too extensive and running too deep to feel, to tinge the sentimental stories I tell myself in outward rituals of lamentation and love. I may weep sputteringly for ghosts, but I cry namelessly, with my whole being, for something that never was. Its fugitive, unknown presence surrounds me: an ocean of mysterious, bottomless romance, inexplicable passions, unspeakable yearnings, fleeting infatuations, intimations of a different life, even another universe.

Someone else did move in from far away.

In an older Westport is an apple tree now lost in the forest where my mother and brother wander, unborn, inside their enchanted glen, unable to escape the spell of forever, to get into this life. I grew up with zombies who stole their bodies and, masquerading as them, enacted a mockery of family.

At this level of intimacy, there is no separating boy from girl, sibling from swain. The dirge I sing is for a darling I never knew.

Even though motorboats populated with reckless teenagers, beer-chugging yuppies, and their rowdy concubines disturb Long Pond, sending waves across it, rocking kayaks and fowl, jack-hammering cams in internal-combustion shafts, the clowns at their helm intentionally running through congregations of sunning gulls and ducks so that they fly up—the luxury liners pass and everything restores itself in a slightly different pattern. The birds resettle with indifference, minor contempt for such inelegant propulsion and useless, bumptious existence. The waves reestablish their pattern: the wind, with one shift in breath, changes the face of the entire lake, gently warping its surface and bending its curves into and against themselves, forming other crisscrossing curves that overlap and vanish instantaneously everywhere, smashing and restoring billions of replica suns. The sound of the boat is swallowed in eternity. Then there is nothing of it except an altered infinity.

Chaos once and forever prevails as Harmony. Pine trees reaching off precipices up Beech Hill and Western Mountain ignore those that have already crumpled over the edge. Others twist out of rocks along sheer walls, making a statement more jagged and defiant than any mere machine. Crickets play dry sticks as they submit to flight. Tiny, almost weightless ants crawl over each boulder world of massive talus slopes deposited by glaciers whose echo dissolves against water-color cirrus stroked in daystar blue.

Here along the Great Notch trail we arrive at a pool for land mammals—irregular slabs of immense flat stone laid atop one another in a giant's stepladder down to the well of creation. Sun pouring through nectar onto underwater megaliths shows a reality more haunting and profound than childhood revealed through a translucent lemon wrapper.

When our generation is gone from the Earth, the rocks at the seawall will be worn away by the waves, their tidepools drained; Thunder Hole will collapse under the sea and become a shapeless crater; Echo Lake will evaporate and fill with whatever "g" forces draw over its basin; the Dipper will bend out of shape, vacating night, all in a single morning and a day, for we will be vanished and unminded, and nothing of ours will impede the acceleration of timelessness, of time. Others, very different, will live on this Earth. My molecules will be among stone and in white birds, if there are white birds then.

Yet this gull's breast glistens whiter now than white as she hovers, scrutinizing the sea.

Manset, Maine, and Kensington, California
October, 2003—August, 2005

Some Incidental Notes

On the Integration of Nature: Post-9/11 Biopolitical Notes follows *Embryogenesis: Species, Gender, and Identity* (North Atlantic Books, 2000) and *Embryos, Galaxies, Sentient Beings: How the Universe Makes Life* (North Atlantic Books, 2003) in a series of texts on biology, consciousness, thermodynamics, and evolution. Although not an embryology text as such, *On the Integration of Nature* is rooted in the biological premises and epistemologies of the prior two books and represents a refinement of them. It also hearkens back to my first "biology" book, which was also my first published book, *Solar Journal: Oecological Sections* (Black Sparrow Press, 1970) in its compilation of dreams, narratives, fragments, scientific metaphors, and essays in a literary, hermetic synthesis.

Page 34, footnote
   The immediate forerunner of all these essays is the chapter "The Wahhabi Critique of Darwinian Materialism" in *Embryos, Galaxies, and Sentient Beings,* pp. 255–280. That piece deals more directly with the covert dialectic between al-Qaeda and biotechnologists.

Page 230
   In Bar Harbor on Monday and Thursdays evenings you can walk out of the early show at Reel Pizza onto the village green to hear the band playing, little kids (and some big ones) running around the bandstand in a giddy circle ritual (including some moms and dads carting their whelps aboard their shoulders),

a round dance probably invented long ago which now is totally accepted and passed down. The band usually closes with "The Star-Spangled Banner." They did that on July 4th, 2005, when Lindy and I attended with her sister and brother-in-law, onetime career-soldier Jim Doyle. He immediately put his hand over his heart. Gazing at the panorama of '50s America, I was moved to tears. I said to Jim, "They can take away the government, but they can't take away the country." Visibly moved too, he said, "Amen."

Page 237

An irony of my life is that, through North Atlantic Books/ Frog, Ltd., I ended up publishing books by both of my rescuers from that night long ago: Alan Powers' *BirdTalk: Conversations with Birds,* 2003, and Sid Schwab's account of his training as a surgeon, tentatively entitled *Cutting Remarks* (2006). Or is this just the way the universe, ignoring time and space, doubles back and synchronizes effortlessly, trying to get us at least to smile at the way it is laughing at itself?

Index

This is a handmade index; no computer was used. I assembled it because: 1. I thought the book needed an index in order to tie its separate parts together. 2. With its (mostly) short separate, disparate pieces, it is not the kind of text that is ordinarily indexed. 3. I wanted an idiosyncratic index that reflected the spirit and content of the book.

The index is unconventional in many ways. The main one is: If you go to the page of an indexed item, you may find only the most glancing or casual reference. This reflects the fragmentary and often elliptical nature of the text itself. My intention was to provide landmarks. The index like the table of contents is a map through the pages and reflects themes and relationships that are often more latent than explicit.

Many categories are inconsistently indexed or not indexed. For instance, I itemized the names of most animals. This was partly for my own amusement (and I regretted its tedium soon enough). I got involved in a zoological inventory because I had created a Noah's ark of creatures. To be less partisan, I threw in an occasional plant, but not all plants. Some things I indexed on an intuition or whim: "apple tree(s)," "balloons," "boat(s)," "catapult," "clock," "egg(s)," "fiesta siesta," "xylophone," "zoo," and colors like "yellow" and "orange."

I indexed almost all place names from Mount Desert Island but only a selection of those from elsewhere because the book is a guide to the regional landscape of Mount Desert, a literary rendition of it. To list all geographical terms from everywhere would have made the index too busy. I did not put (Mount Desert) beside its locations.

I took a stab at indexing concepts that are not easy to define in a single word or pin down objectively: "unfathomability," "bad," "here," "there," "no one," "whatever," "opposites" ("antitheses"), "existence feeling unreal to itself," "distinction between incommensurate forms," etc.

I tried to hold most indexed categories in my head, but invariably some are more fully inventoried in sections of the book. Discovering abandoned items during later passes, I have left them in, even though the index would be cleaner without them.

At times, as certain terms overlap, one set of categories gradually replaces another: "existence" and "creation"; "life" and "biology"; "existence" and "reality"; "meaning" and "reality"; "illusion," "porosity of matter," and "hallucination";

"creation" and "life"; "matter as energy" and "energy as matter"; "extraterrestrial life," "alien life," and "cryptozoology"; "nothing," "is-not," and "nihilism"; "identity" and "ego"; etc.

Authors are provided for books but not films, except where the director and/or author is otherwise discussed in the text.

If an item is not anywhere on a page, try the preceding or succeeding page. Sometimes the hand is faster than the mind, and sometimes it is the other way around.

I have used the following code:

The conventional f. for an item that appears in a footnote, e.g. 12f.

My own invented (a.), meaning "by allusion," for an item not identified in the text, e.g. 12(a.). If the item is a subject matter, the (a.) indicates that it is discussed but not named. If an author, he or she is the source of ideas on the page. This usage is not exhaustive in that, as per most indexes, broad concepts are often indexed without an (a.), though they *do not* appear by name on the page.

My own invented (m.) to indicate an entry that appears not as a real thing (or natural animal) but a metaphorical or mythological version of itself.

on the bridge in Bangor, Maine; photograph by Lindy Hough

A graduate of Amherst College, Richard Grossinger received a Ph.D. in anthropology from the University of Michigan, writing an ethnography of fishing in Maine. He is the author of many books, including *Planet Medicine; The Night Sky; Embryogenesis: Species, Gender, and Identity; Embryos, Galaxies, and Sentient Beings: How the Universe Makes Life; Homeopathy: The Great Riddle; New Moon;* and *Out of Babylon: Ghosts of Grossinger's.* He and his wife Lindy Hough are the founding publishers of North Atlantic Books in Berkeley, California.